The Shrub Identification Book

THE
Shrub Identification
BOOK

*The Visual Method for the
Practical Identification of Shrubs,
Including Woody Vines and Ground Covers*

By

GEORGE W. D. SYMONDS

Photographs by A. W. Merwin

Quill

A HarperResource Book
An Imprint of HarperCollinsPublishers

THE SHRUB IDENTIFICATION BOOK. Copyright © 1963 by George W.D. Symonds. All rights reserved. Printed in the United States of America. No part of this book may be used or reproduced in any manner whatsoever without written permission except in the case of brief quotations embodied in critical articles and reviews. For information address HarperCollins Publishers Inc., 10 East 53rd Street, New York, NY 10022.

HarperCollins books may be purchased for educational, business, or sales promotional use. For information please write: Special Markets Department, HarperCollins Publishers Inc., 10 East 53rd Street, New York, NY 10022.

Library of Congress Catalog Card Number 63-7388

ISBN 0-688-050409

02 03 04 05 06 QK 20 19 18 17 16 15 14 13

To:
ROBERT
GEORGE
DAVID
FRANCES

ACKNOWLEDGMENTS

The photographs in this book are all by A. W. Merwin of Wilton, Connecticut, and speak for him better than I can. I want to add, nevertheless, that in addition to his splendid photography, his sustained interest and patience under difficult conditions are greatly appreciated.

The assistance given to me by Mr. E. J. Alexander, of the New York Botanical Garden, both in the field and during many hours of consultation, was invaluable. I can only thank him for his generous and skillful guidance and say that I have seldom met anyone with such extensive knowledge hidden beneath such an unassuming and charming exterior.

I particularly wish to thank Senator Thomas C. Desmond of New York State for his kindness in giving me complete freedom to use his fine arboretum in Newburgh, N. Y. His superintendent, Mr. Rudolph Nocker, was also most helpful. The fine collection of native plants in this arboretum is especially outstanding.

It is impossible for me to thank all the people adequately who assisted me in many ways, but I do want to thank the following not only for their assistance in locating various shrubs, but especially for taking me to rare and wonderful places:

Professor Earl L. Core, Chairman, Department of Biology, West Virginia University

Mr. Frederic L. Steele, Tamworth, New Hampshire

Mr. & Mrs. H. Lincoln Foster, Falls Village, Connecticut

Mr. E. C. Childs, Norfolk, Connecticut

Mr. G. G. Nearing, Ramsey, New Jersey

Mr. Alton Miller, Mr. Ray Dubreuil, and Mr. Patrick Deming of the Connecticut Agricultural Experiment Station, Canaan, Connecticut

Mr. Stanley J. Smith, Curator of Botany, New York State Museum, Albany, New York

I also wish to thank the following for sending me fresh specimens:

Professor H. R. Totten, University of North Carolina

Mrs. Barbara McKenzie, Elgood, West Virginia

Mr. C. A. Eulett, Otway, Ohio

Dr. Gertrude E. Douglas, Feura Bush, New York

Mr. Kenton L. Chambers, Curator, Herbarium, Oregon State College

Dr. A. J. Sharp, University of Tennessee

For special assistance in technical matters, I thank:

    Dr. Edward G. Voss, University of Michigan

    Professor Oswald Tippo, Provost, University of Colorado, formerly Chairman, Botany Department, Yale University

    Professor John R. Reeder, Botany Department, Yale University

    Dr. Arthur Harmount Graves, Wallingford, Connecticut

    Mr. Leonard J. Bradley, Staff Botanist, Audubon Center, Greenwich, Connecticut

Also thanks go to the following for their assistance in locating plants in botanical gardens:

    Mr. L. P. Politi and Miss Bride McSweeney, New York Botanical Garden

    Mr. George Kalmbacher, Brooklyn Botanic Garden

    Dr. Donald Wyman, horticulturist, Arnold Arboretum, Jamaica Plain, Boston, Massachusetts

    Mr. Henry Draper, Case Estate, Arnold Arboretum

CONTENTS

INTRODUCTION

HOW TO USE THIS BOOK

PART I—PICTORIAL KEYS for Broad-leaved Upright Shrubs

    Key #1 Thorns
    Key #2 Leaves
    Key #3 Flowers
    Key #4 Fruit
    Key #5 Twigs
    Key #6 Bark

PART II—MASTER PAGES 1-200

    Broad-leaved Upright Shrubs, pp. 1-156
    Needle-leaved Shrubs (Upright and Prostrate), pp. 157-162
    Broad-leaved Ground Covers, pp. 163-171
    Vines (all Broad-leaved), pp. 172-200

BIBLIOGRAPHY

GLOSSARY

INDEX TO MASTER PAGES

THORNS—TH

LEAVES—LF
Opposite—Opp

FLOWERS—FL

FRUIT—FR

TWIGS—TW
Opposite—Opp

BARK—BK
Opposite—Opp

MASTER PAGES—MP
Needle-leaved—NE
Ground Covers—GC
Vines—V

INTRODUCTION

This is an age of paradox and enigma:

It is an age of conservation; it is an age of wild spending of resources; an age of speed, but also an age of leisure.

It is an age for city dwellers, who travel and explore; an age of multitudes and noise, but solitude and silence are needed as never before.

This is an age of splendid sights, but it is an age beset with blindness; an age full of promise and opportunity, yet an age courting final disaster.

This is an age for greatness, but an age when simplicity is sometimes mistaken for smallness.

This is an age on a threshold of many vistas, but an age when few vistas are clear.

On this threshold, one of the paths is clear and simple. It is well trodden, but too often neglected. This is the path of simple pleasures that have always been there. Among them are the wonders of nature. This is the land free, to be used for joy and gladness, for freedom of the body and the mind. There is no bit of land so poor that discovery can be denied to those who seek it.

It has been my privilege to walk with a man who knew lichens and mosses, along a road ugly to the untrained eye. This man made the dullest-looking border become alive, full of exciting new possibilities, the squalid surroundings forgotten amidst such splendor. Here before one's eyes a miracle happened, wonders falling one over the other on the expanse of what had shortly before appeared to be only dirt, rocks and a few dusty plants. Here were wonders unexcelled in the most magnificent garden.

This time of excitement was all too short, a once in a lifetime experience. From it was gained an awareness that made possible a continuing appreciation of the less evident glories of nature. All that is required is a willingness to see what is there and a desire to understand what is seen. It is not a difficult step, but the step must be taken. The woods, the fields, the parks and wilderness areas of mountain and plain are there to be used. Even a park in the largest city has more on every hand than the largest museum. If on a walk, one really looks at the trees, shrubs, flowers, ferns, mosses and lichens, the numbers and variety are truly overwhelming until some order is made of this seeming chaos. If, then, one first picks out a few plants that are known, a start has been made, and with interest aroused, a truly fascinating pursuit begins, the thrill of personal discovery becoming real. The problems involved in identifying plants are sufficiently difficult to create a challenge, one as good as any puzzle, but more rewarding, as the solution is not the end.

In taking up a study of plants, it is perhaps best to begin with trees and shrubs as they are the easiest to identify. Success is necessary; bewilderment and failure are fatal. The goal of identification is recognition. To recognize means to re-know, and to see a familiar plant, especially in an unfamiliar setting, is indeed a thrill. This knowledge, once gained, is always at hand, and with plants available, one can never be bored.

Thus, although this book is designed specifically as a practical method for successfully identifying native and naturalized shrubs in the area covered, there should be a double incentive for using it: the pleasures of recognizing plants far outweighing the immediate objective of identifying them.

The method for shrub identification used in this book is based on the use of photographs of actual details of the plants involved. The section, "How To Use This Book," explains the process of using these details to determine the identity of any particular shrub. At this point, however, a brief explanation of botanical identification is given to avoid future misunderstanding or disappointment. As the world of plants, like that of people, is made up of individuals, it should be apparent that no two shrubs are ever exactly alike. There are, nevertheless, certain characteristics possessed by every shrub which indicate its relationship to other shrubs. For example every Lilac has a number of features that are similar to those of all Lilacs and which distinguish it from other shrubs. It is by a deliberate and directed use of these details that identification is accomplished, not only by the botanist but by anyone attempting to do so.

It sometimes happens that some of the details vary among plants of the same kind, which is confusing to the novice; but despite these apparent discrepancies, all plants do exhibit a pattern and consistency of detail which makes classification always possible. Only the matter of degree is the stumbling block. For example, great variations may occasionally be encountered, but, as there is always a cause, logical reasons can usually be found to account for them. All plants have evolved over long periods of time, and when the environment has remained relatively constant long enough, stable forms are established. When the environment has not remained constant, for one reason or another, plants may not achieve (or retain) a stable form. This condition is particularly evident among certain kinds of plants that have been directly affected by extensive clearing of land in this country. These exceptions to the standard form are indicated, but few, if any, specific names are given for such plants in this book. It should be understood that it is not necessary to attempt exact identification of a shrub under these conditions, but rather to place it in the small group of related plants to which it is found to belong. This is almost always a practical and correct solution, infinitely more correct than trying to force an exact (named) identity on a plant when none exists. The great majority of shrubs are stable enough so that precise identification is usually possible. Final identification in this book, except for the unstable forms mentioned above, means reducing the possibilities to the correct species, such as Mountain Laurel or Flowering Dogwood. The botanist carries the process further into varieties (*Pink* Flowering Dogwood, for example), hybrids (crosses between species) and specialized forms, but this is not generally necessary for practical purposes, or, as in the case of the Pink Flowering Dogwood, the variation from the standard is obvious.

The shrubs included in this book grow within an area bounded roughly by the eastern half of the United States and Canada. As this book is not a complete manual, certain plants were omitted so that unnecessary numbers would not interfere with the main purpose. Records of all woody plants known to be growing in the area were studied and compared with reports obtained from over twenty states. From this study, a pattern of distribution and frequency emerged for each shrub, which proved to be a valuable guide in determining which shrubs to include. The omissions fall mostly into the following categories:

(1) Shrubs from other regions that overlap small sections of the area covered by this book.

(2) Strictly southern or subtropical plants.

(3) Foreign shrubs which, although planted here, have not become naturalized to any extent in this country. (A number of foreign shrubs have become established here and now appear to be almost "native." These are included.)

(4) Hybrids, varieties and unusual forms.

(5) Some shrubs belonging to genera whose species are very numerous. It was felt in these cases that a number of the most important and representative species would be sufficient, for once the genus is known, identification of any additional species can be made with the help of a botanical manual, available at most libraries, or, usually, by consulting a list describing the local flora. The Hawthorns have hundreds of species, but the specialists (people who may spend years on this genus alone) never agree as to just how many there are or when to call a particular plant a real species rather than only a variety of another. In this case, only the genus (Hawthorn) is given here, with the warning that naming of species is better left to others.

Both popular and botanical names are given for each shrub, the botanical ones in accordance with Alfred Rehder's *Manual of Cultivated Trees and Shrubs,* Macmillan, second edition. There are two exceptions to this: the Clubmosses (*Lycopodium*) and Japanese Knotweed (*Polygonum cuspidatum*). These are not included in Rehder's manual, and the botanical names follow the eighth edition of *Gray's Manual of Botany,* by M. L. Fernald.

HOW TO USE THIS BOOK

The objective in any but the most casual system of identification is the classification as well as the naming of something. In one way or another, the method for doing this is based on the elimination of possibilities. This usually involves the reduction of large numbers of such possibilities to fewer and finally to one. Plant identification is based on plant relationships. Botanical names (and some common or popular names) are also based on these relationships. For practical purposes, there are two terms to keep in mind: (1) genus and (2) species. *Genus* is applied to a closely related group of plants, such as the Dogwoods or the Honeysuckles. The *species* is the particular kind of Dogwood or Honeysuckle. The botanist places the genus name first, with a capital letter, followed by a specific epithet in small letters, thus: *Cornus florida*—Flowering Dogwood (*Cornus*—Dogwood; *florida*—Flowering). Botanical nomenclature has several advantages: It is understood in all countries, is based on natural relationships of plants and is consistent in form. Common names, on the other hand, were not originated with these features in mind. However, as botanical names are unfamiliar to many people and appear somewhat as mumbo-jumbo to the uninitiated, common names have been used throughout this book. The disadvantage of doing this is, of course, that the particular name selected is not always the best known in all sections of the country. To overcome this, alternative popular names (together with the botanical *one* for each shrub) are given in the main section of the book devoted to information concerning individual plants, with an apology made here for the omission of someone's favorite.

The method used in this book to identify shrubs utilizes photographs of specific parts of each shrub. By placing similar details, such as leaves, together in separate groups, the groups are quickly found, and the differences between differentiating details within each group are easily noted, thus eliminating all but a few possibilities. Finally, by comparing these few with a few possibilities within an entirely different group (or groups), all shrubs but the correct one are eliminated.

There are five main parts (or details) of a shrub to look for: Leaves, Flowers, Fruit, Twigs and Bark. Each of these has a number of smaller details which are pointed out in the book, all of which can easily be seen in the photographs. All of the five major details may not be available at any one season of the year, but usually a combination of any two or three is sufficient for practical purposes. In the very few cases when this is not so, it may be necessary to wait for an additional one. This should not be a source of discouragement when one realizes that a botanist might have to wait also. Plants are not machine-made products, but living, changing things, and one must not expect that each individual shrub will be exactly like every other. One can expect, however, that the individuals grouped together under one specific name will be sufficiently consistent in the total combination of their details to make identification reasonably certain.

A shrub is usually defined as a relatively small, woody plant with several stems (instead of a single trunk as with most trees). It might be simpler to think of a shrub as any woody plant that is not a tree, as the terms "tree" and "shrub" are actually used only as a convenient means of dividing woody plants into two easily recognizable groups. For further convenience in reducing large numbers to smaller, shrubs have been divided here into four readily distinguishable groups: (1) Broad-leaved upright shrubs, (2) Needle-leaved shrubs of all kinds, (3) Broad-leaved ground covers, and (4) Vines. The plants in each of these groups are in separate sections of Part II of this book, called the Master Pages. The broad-leaved *upright* shrubs constitute by far the largest and therefore the most difficult group to identify, and for this reason the first part of this book is in the form of Pictorial Keys which are designed for the identification of the *genus* of these plants. The last step, *species* identification, is accomplished in the second part of the book. When a plant belongs to one of the last three groups above, reference should be made directly to the appropriate section of the Master Pages. As there are relatively few of these plants, no special Key has been given for them. They are all distinctive and can be found quickly within the few pages of the sections involved. Before using this book, the Introductions to these three sections should be consulted to see just what types of plants are included in each of them.

There are six Keys for genus identification of the broad-leaved *upright* shrubs. In addition to a Key for each of the five major details mentioned previously, a Key for thorns is given first, as it is a quick guide when thorns are present. The other five Keys are the basic ones, however. Note that the contents page of this book is indexed at the edge to correspond to similarly placed markings on the outer edges of the pages of the book, making it easy to find a particular Key or Master Page section. Each Key, except for the Thorn one, has a short Introduction explaining how it is arranged. These should be read before using the Keys for the first time; thereafter they will be found useful as checklists of things to look for.

When using the Keys, start with the first one and look for thorns on the actual plant, then continue through the Keys in the order in which they are presented, until satisfied that the genus has been correctly determined. Do not start with Key #6, Bark, for example, as this Key is designed only to verify identification made from previous Keys. Also, do not dismiss leaves or fruit in winter, as they may hang on for much of the year. Twigs, although primarily a winter characteristic, are often very helpful in summer, the buds often being well developed before the leaves fall off. It will become apparent with actual use that a very good idea of identity is obtained from the first applicable Key; however, several Keys should be used to avoid jumping to faulty conclusions, and will be found essential for the plants that are difficult to identify. If, for example, it is difficult to decide between the pictures of three different twigs, turn to the Bark Key, and the chances are good that the bark of the actual plant will be like only one of the three possible plants selected in the Twig Key. Emphasis is again made that final identification should never be based on one detail alone. Any specific detail of an individual plant of any one species may vary considerably from the average for the species, but the combination of all available details will be found to be indicative for that plant and species. (In cases when this is not found to be so, the plants will almost certainly prove to be varieties, hybrids or other unusual forms.)

A very important feature to keep in mind is the Alternate and Opposite characteristics of shrubs. The leaves, buds and leaf-scars (the places marking where leaves separated from the twigs when they dropped off in the fall) of every plant are arranged either *alternately* along the stems, or in *opposite* pairs (occasionally in whorls of three to four around the stem at definite points called nodes). This is a consistent characteristic of almost all woody plants. The Leaf, Twig and Bark Keys have been subdivided into Alternate and Opposite sections, and the first step in identifying a shrub is to notice whether it is alternate or opposite. Two shrubs that are exceptions to the rule commonly have both alternate and opposite leaves and buds on the same plant and are represented in both sections of the Keys. A few shrubs that are actually alternate produce leaves in clusters at the ends of the twigs, giving an opposite impression. These will be found in both sections of the Leaf and Bark Keys, but only in the Alternate section of the Twig Key, as their buds are obviously alternate.

Under each picture in the Keys is a name and a number. When going from one Key to the next, always look for the correct picture and not for the name selected in the previous Key, as several different pictures are frequently given for the same genus. Only when the picture corresponds to the actual detail should the name correspond to the one previously selected, thus verifying identification. When the genus has been determined, turn to the Master Page indicated by the number shown next to the correct name. The Master Pages bring together all the details of the individual shrubs. Whenever a particular genus, such as Dogwood, includes two or more species, the corresponding details of the different species are shown together, so that final (species) identification is made by contrasting these different details. If the genus of a particular shrub is already known, it is not necessary to use the Keys, and by consulting the index in the back of the book, the correct Master Page can be turned to directly for species identification.

The pictures in this book are all actual size except in a few cases, and then the scale is indicated. Details that are to be compared are always in the same scale.

A small hand lens is useful and interesting to use, but is not necessary except in cases where very small details must be clearly seen, and this mostly to distinguish between very similar varieties or forms of the same species, or for other specialized work.

PART I

PICTORIAL KEYS

BROAD-LEAVED UPRIGHT SHRUBS
(all actual size except when noted)

CONTENTS: Key #1 Thorns, Prickles, Bristles and Similar Growth

Key #2—Leaves

Key #3—Flowers

Key #4—Fruit

Key #5—Twigs

Key #6—Bark

KEY #1 THORNS, PRICKLES, BRISTLES
and Similar Growth

The leaves, buds and twigs of the three plants in this top row are found opposite one another; all other thorny plants in this book produce alternate growth.

very twiggy but not actually thorny

Twigs often end in sharp points; the twigs and leaves are silvery.

Buckthorn 84

Viburnum 136

Buffalo-berry 95

No actual leaves are produced, only leaf-stalks, which are sharp and thorn-like.

evergreen

Gorse 67B

Firethorn 42

Plum 54

Oleaster 96

Hawthorn 43

Sarsaparilla 92

small stems and twigs round;
green, red or maroon

Thorns

buds usually small; shiny
red (or green)

rust-
colored
buds

Rose 57

Prickly Ash 70

stems hardly woody, with very large pith;
often angled (not round); green, red or maroon

Remnants of leaf-stalks (below bud)
often persist on stems.

**Bramble 46
(Blackberry)**

Stems are actually red, but almost always covered by a heavy bloom that makes them look white or pinkish.

**Bramble 46
(Raspberry)**

small stems and twigs round; green, red or maroon; buds usually small, shiny red (or green)

small stems and twigs round; green, red or maroon

Rose 57

Rose 57

Gooseberry 30

Barberry 18

Gooseberry 30

Matrimony-vine 125B

rough bark;
dark gray
or brown

small stems and twigs round; green,
maroon or red; buds usually small,
shiny red (or green)

stems often angled (not round); green,
maroon or red; remnants of leaf-stalks
often persist on stems into winter

Rose 57

Bramble 46
(Blackberry)

Barberry 18

Gooseberry 30

Thorns and Bristles

bristly

stems densely
red-hairy
(not bristly)

Rose 57

Bramble 46

bark dark gray

Gooseberry 30

Rose 57

stems golden brown (or red),
with or without bloom

small stems and twigs
green, red or maroon

twigs often bristly

buds shiny
red (or green)

Bramble 46
(Raspberry)

Rose 57

bark of larger
stems dark gray

bristles
rust colored

stems, twigs and
bristles golden brown

Bristly Locust 69

Currant 30

Sarsaparilla 92

Key #2—LEAVES

This Key is arranged as follows:

I ALTERNATE LEAVES, pp. 1-25

    A Simple Leaves—one main leaf blade on leaf-stalk (petiole)

        (1) With teeth evident along edges of leaf, pp. 1-11

        (2) Without teeth (entire) or with indistinct teeth, pp. 12-19

    B Compound Leaves—with more than one leaflet on leaf-stalk, pp. 20-25

II OPPOSITE LEAVES, pp. 26-36

    A Compound Leaves, p. 26

    B Simple Leaves

        (1) With teeth, pp. 27-30

        (2) Without teeth or with indistinct teeth, pp. 31-36

Note: Some shrubs which actually have alternate characteristics often produce leaves in clusters at the ends of the twigs, giving an opposite effect. These leaves will be found in both the alternate and opposite sections. Two shrubs, Common Buckthorn and Fringe Tree, although predominantly opposite, often have some alternate leaves and these appear in both sections. Shrubs which often have different types of leaves are represented on more than one page.

The simple leaves in each group are arranged with the largest leaves first and with the very small or very narrow ones last.

In comparing actual leaves with the photographs, look particularly for the following five characteristics:

(1) EDGE OF LEAF: if with teeth, note differences in size, shape, and direction (whether pointing outward or forward toward the end of the leaf); also whether singly or doubly toothed (small teeth on larger ones); also whether the teeth are all the same size and evenly or unevenly spaced. Sometimes there are no regular teeth but the leaf is wavy-edged. A few shrubs have leaves both with and without teeth, in which case they are shown in two or more places in the Keys.

(2) TIP OF LEAF: long-tapered, short-pointed, sharp-pointed, blunt or rounded.

(3) BASE OF LEAF: tapered, rounded, straight across or heart-shaped

(4) LEAF-STALK (petiole): long, short, wide or narrow

(5) VEINS: arrangement—large (primary) veins branching off along whole length of midrib (pinnately), or starting mostly at base of leaf from approximately one point (palmately); also note whether the main veins are straight and go directly to the edges of the leaf, or whether they branch before reaching the edges; also note if veins curve upward and whether they are prominent or indistinct.

A brief examination of this Key will indicate that leaves which appear to be superficially alike will almost always have at least one of the above characteristics markedly different. Be sure when using this Key that the differences noted are clear-cut. For example, if the leaves of a particular shrub have very short leaf-stalks it would be safe to assume that leaves of all shrubs of the same kind would also have short leaf-stalks and would be found so in the Key pictures. Do not, on the other hand, try to distinguish between leaves based on leaf-stalks in the middle range. Do not base final identification on leaf size, as leaves vary greatly in this respect. In fact, it should be noted that, in collecting the leaves for this Key, a deliberate effort was made to show all similar leaves in the same size so that variations of size would not influence comparisons. The five details listed above are the ones to concentrate on after the general average size is determined. Occasionally, a note is given of other characteristics as, for example, if leaves are distinctly whitened beneath or covered with evident dots, this is indicated. Otherwise, it can be assumed that the five points are sufficient to distinguish between the leaves shown.

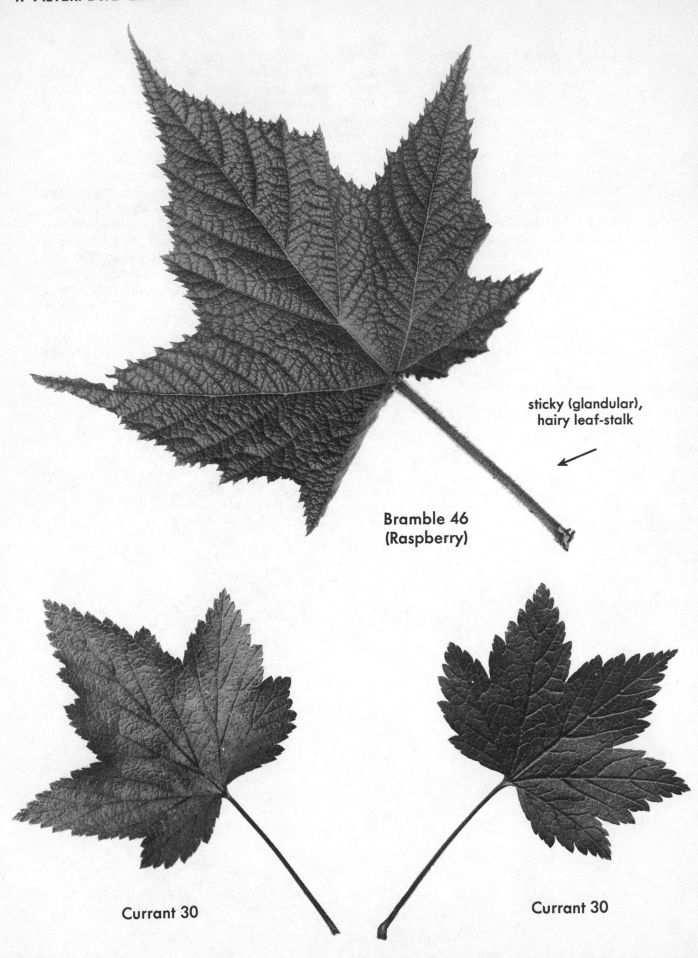

sticky (glandular),
hairy leaf-stalk

Bramble 46
(Raspberry)

Currant 30

Currant 30

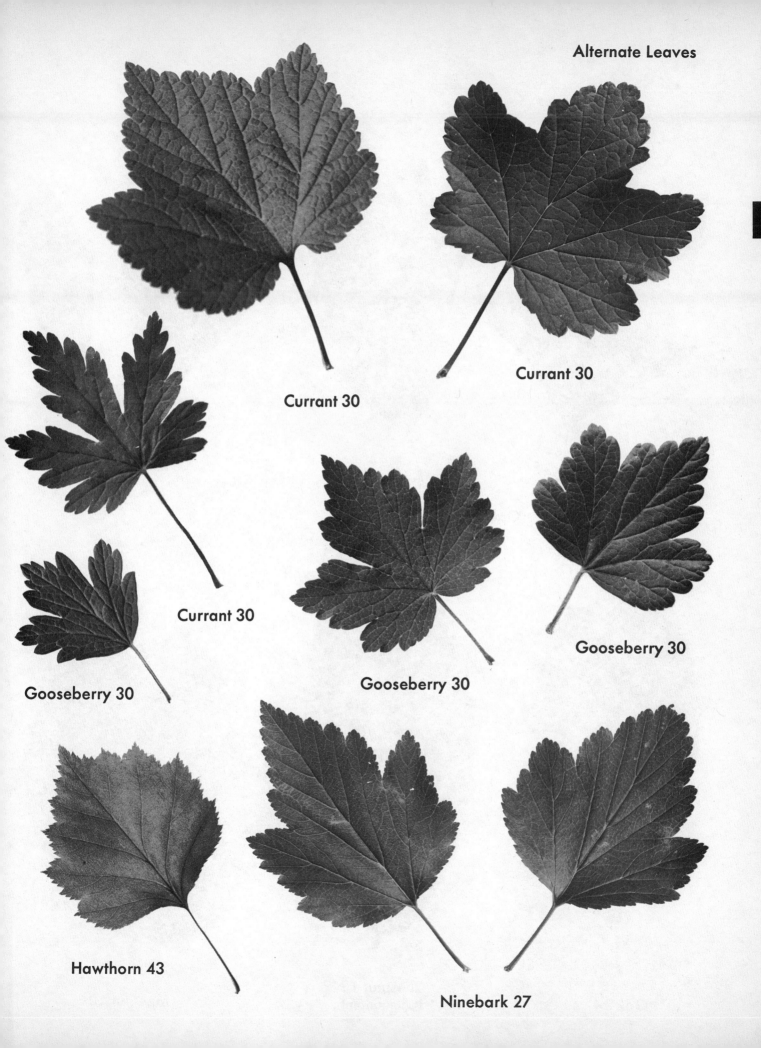

Currant 30

Currant 30

Currant 30

Gooseberry 30

Gooseberry 30

Gooseberry 30

Hawthorn 43

Ninebark 27

Oak 14

Groundsel-bush 155

Oak 14

Chestnut 13
(Chinquapin)

Hazelnut 10

Rose of Sharon 90

Witch-Hazel 38

Silverbell 128

Alder 6

Holly 78

Alder 6

Alternate Leaves

Hazelnut 10

Alder 6

Alder 6

Hornbeam 12

Shadbush 44

Spirea 40

Birch 9

LF
6

Plum 54   Buckthorn 84   Plum 54

Clethra 102   Holly 78

Alternate Leaves

LF
7

New Jersey Tea 45A

Virginia-willow 23B

Leucothoë 118

usually
opposite,
occasionally
partly alternate

Buckthorn 84

Alder 6

Holly 78

Buckthorn 84

Holly 78

Leucothoë 118

Willow 1

Cherry 54

Note: There is a row of small
black hair-like glands
on the mid-rib, visible
to the eye but best
seen with a lens.

Shadbush 44

Chokeberry 39

This row is all Pussy Willow 1.

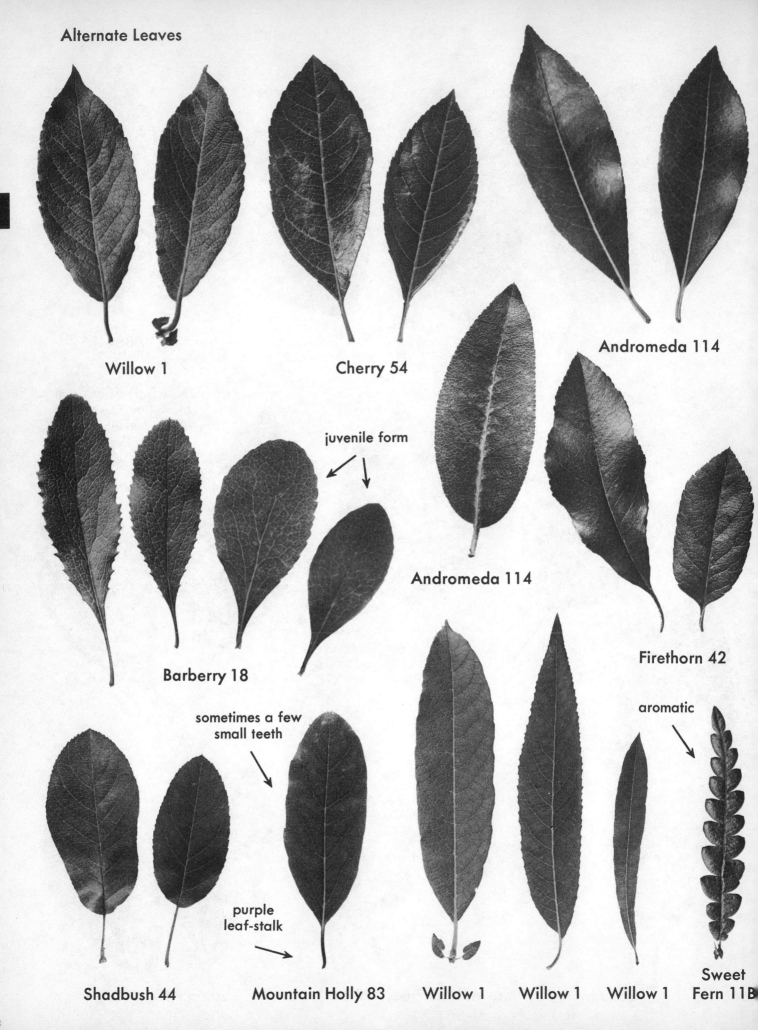

Alternate Leaves

Willow 1

Cherry 54

Andromeda 114

juvenile form

Andromeda 114

Barberry 18

Firethorn 42

sometimes a few
small teeth

aromatic

purple
leaf-stalk

Shadbush 44

Mountain Holly 83

Willow 1

Willow 1

Willow 1

Sweet
Fern 11B

aromatic

aromatic

Bayberry 4

Bayberry 4

Groundsel-bush 155

Holly 78

aromatic

Wax-myrtle 4

Barberry 18

Alternate Leaves

aromatic

Sweet Gale 4

Spirea 40

Hawthorn 43

upper leaves usually alternate; lower ones opposite

Marsh-elder 156

Spirea 40

Birch 9

leaves covered with many small dots

Leatherleaf 116B

pale, often rusty-woolly, beneath

Spirea 40

Barberry 18

Blueberry 119
(Bilberry)

Huckleberry 126

LF
12

Knotweed 17B

Smoke-tree 72

Redbud 68

Rhododendron 104

Rhododendron 104

Alternate Leaves without Teeth

mostly opposite, but occasionally partly alternate

aromatic

Fringe-tree 129

Spice-bush 21

small teeth not always evident

veins particularly conspicuous early in the season

Leucothoë 118

Buffalo-nut 16

usually with teeth, but edges often curl down, hiding them

LF 14

Smoke-tree 72

Alder 6

Dogwood 98 (Alternate-leaved)

Buckthorn 84

Lyonia 117

Leatherwood 91A

Alternate Leaves without Teeth

dark green above,
lighter green beneath

dark green above,
usually rusty beneath

Rhododendron 104

short, abrupt
tip (mucro)

silvery
beneath

Laurel 110

sometimes
a few
indistinct
teeth

purple
leaf-stalk

often hairy along
edges (ciliate)
(see also pp. 16, 17)

Mountain Holly 83

Oleaster 96

usually some teeth
(see also p. 8)

Azalea 104

Pussy Willow 1

often hairy along edges (ciliate) (see also pp. 15, 17)

Lyonia 117

Azalea 104

rough (to touch) along edges, but no regular teeth
(see also p. 17)

(see also p. 14)

Leatherwood 91A

Lyonia 117

This row is all Blueberry 119.

Alternate Leaves without Teeth

usually some teeth
(see also p. 8)

short, abrupt
tip (mucro)

purple
leaf-stalk

Pussy Willow 1

Mountain Holly 83

Andromeda 11

often hairy along
edges (ciliate) (see also pp. 15, 16)

inconspicuous
teeth

bluish cast
above, pale
beneath

Azalea 104

Andromeda 114

Huckleberry 126

rough (to touch) along edges,
but no regular teeth (see also p. 16)

bluish cast
above, pale
beneath

Lyonia 117

**Blueberry 119
(Deerberry)**

Huckleberry 126

light green above, pale or whitish beneath

bluish cast above, pale beneath

aromatic

LF 18

Laurel 110

Azalea 104

Bayberry 4

Note: Narrow-leaved form of European Buckthorn, not shown in the Master Pages.

Daphne 91B

Buckthorn 84

Matrimony-vine 125B

Willow 1

dull green above, silvery beneath

aromatic

underside of leaves densely woolly; wool green at first, later in season rusty orange; edges of leaves curl under (revolute)

Oleaster 96

Wax-myrtle 4

Labrador Tea 116A

Willow 1

Alternate Leaves without Teeth

distinctive whitish tip (mucro)

pale beneath; scattered whitish hairs on upper side and along leaf edges

very small leaves on fruiting branches

bluish cast above, pale beneath

leaves at ends of stems very small

leaves covered with many small dots

Blueberry 119 (Deerberry)

Leatherleaf 116B

Menziesia 103

short, abrupt tip (mucro)

short, abrupt tip (mucro)

leaves at ends of stems very small

Huckleberry 126

Blueberry 119

Azalea 104

silvery both sides; brown-dotted beneath

Dyer's Greenweed 65A

Oleaster 96

bluish green

bluish cast above, whitish beneath

Barberry 18

Blueberry 119 (Bilberry)

Sand-myrtle 113B

Bog Rosemary 113

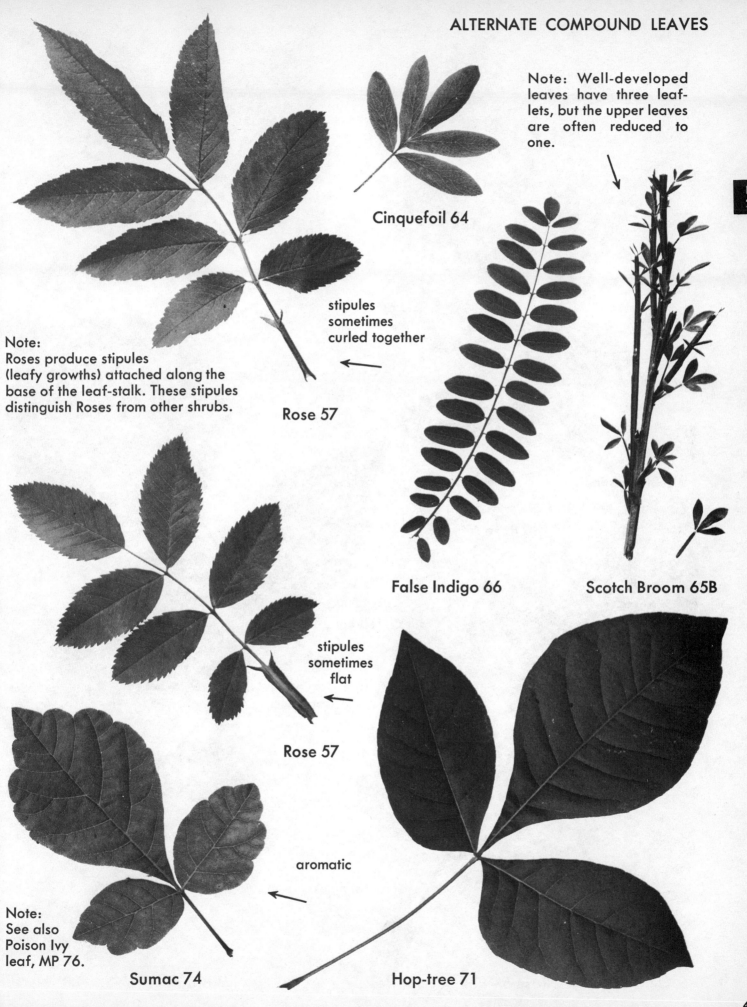

Note: Well-developed leaves have three leaflets, but the upper leaves are often reduced to one.

Cinquefoil 64

stipules sometimes curled together

Note:
Roses produce stipules (leafy growths) attached along the base of the leaf-stalk. These stipules distinguish Roses from other shrubs.

Rose 57

False Indigo 66

Scotch Broom 65B

stipules sometimes flat

Rose 57

aromatic

Note:
See also Poison Ivy leaf, MP 76.

Sumac 74

Hop-tree 71

49

Alternate Compound Leaves

stipules

Rose 57

feathery stipules

Rose 57

leaf-stalk
sticky, red-hairy

Bramble 46

stipules

Rose 57

Bramble 46

LF
22

This page is all Brambles 46
(Blackberry and Raspberry).

Alternate Compound Leaves

This page ½
actual size

Note wings on main
leaf-stalk.

no wings
no teeth

small teeth along
edges of leaflets

Dwarf Sumac 74
(non-poisonous)

Poison Sumac 74
(poisonous; worse
than Poison Ivy
for many people)

Prickly Ash 70

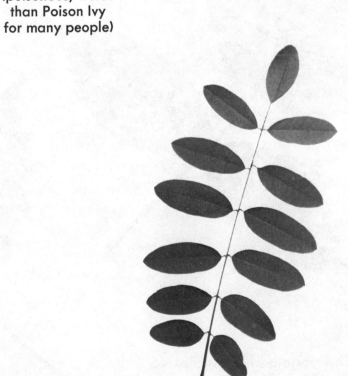

Bristly Locust 69

False Indigo 66

Sarsaparilla 92

Shrub Yellow-root 25

single teeth
along edges
of leaflets

doubly serrate
(small teeth on
large ones)

sometimes
very large

Sumac 74

False Spirea 24

Alternate Compound Leaves

½ actual size

This is all one compound leaf; these leaves are often 3 feet in length.

Sarsaparilla 92

II. OPPOSITE LEAVES

(all are actual size)

Elderberry 134

Bladdernut 87

Forsythia 131

juvenile form
of leaf often
found on new
shoots or young
growth

All leaves on this page are Viburnum 136.

Viburnum 136

Jetbead 45B

Viburnum 136

Mock-orange 22

Hydrangea 28

red
leaf-stalk ←

reddish
or pinkish
leaf-stalk →

Hydrangea 28

Viburnum 136

Viburnum 136

often
wavy-edged
leaf-stalk →

Bush-honeysuckle 154

Euonymus 80

Viburnum 136

Euonymus 80

leaf wavy-edged
with or without small
rounded teeth;
margin often
curls under

Euonymus 80

Deutzia 26

Viburnum 136

These two mostly without teeth
see also   see also
p. 35   p. 36

Coralberry 145

Buckthorn 84

Snowberry 145

Euonymus 80

lower leaves
opposite, upper
ones largely
alternate

mostly alternate,
rarely partly
opposite

Viburnum 136
Note: This is a narrow-
leaved form of Nanny-
berry, not shown in the
Master Pages.

Bush-honeysuckle 154   Forsythia 131   Marsh-elder 156   Andromeda 114

OPPOSITE LEAVES WITHOUT TEETH (entire) or with Indistinct Teeth

top side rough to touch when rubbed from tip toward base; aromatic when crushed

Dogwood 98

Sweet Shrub 20

leaves clustered at ends of twigs

Rhododendron 104

Rhododendron 104

Viburnum 136

mostly in whorls of
three or four around stem
at each joint

Buttonbush 144

Dogwood 98

Dogwood 98

Fringe-tree 129

Opposite Leaves without Teeth

This row is all Dogwood 98.

Mock-orange 22

Lilac 130

Honeysuckle 146

Honeysuckle 146

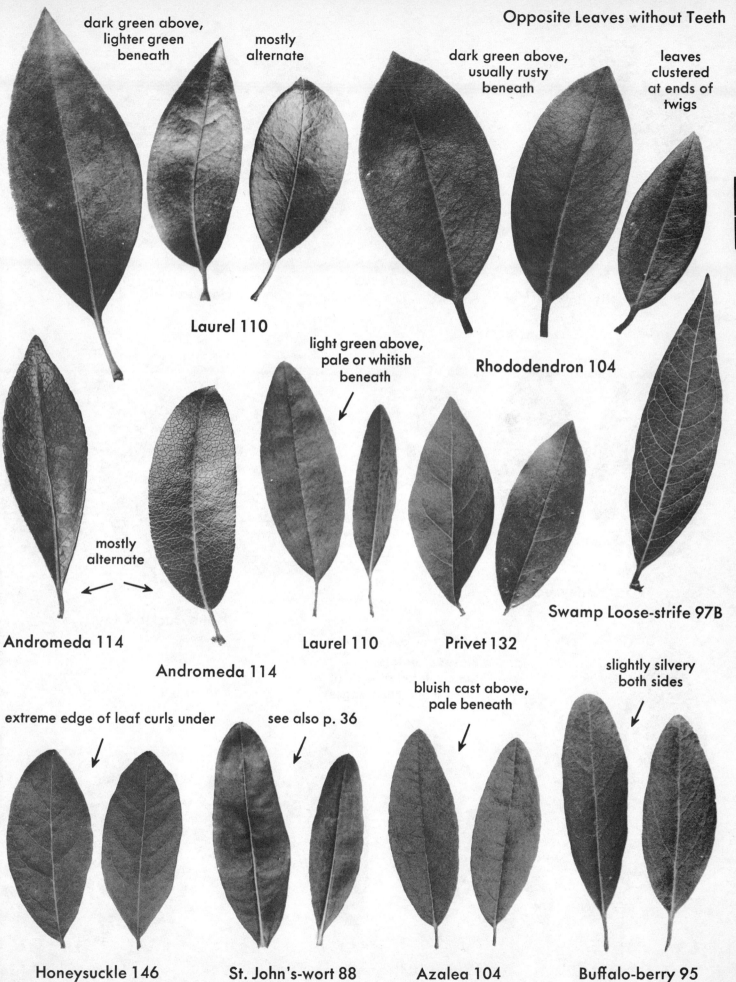

dark green above, lighter green beneath

mostly alternate

dark green above, usually rusty beneath

leaves clustered at ends of twigs

Laurel 110

light green above, pale or whitish beneath

Rhododendron 104

mostly alternate

Andromeda 114

Andromeda 114

Laurel 110

Privet 132

Swamp Loose-strife 97B

slightly silvery both sides

bluish cast above, pale beneath

extreme edge of leaf curls under

see also p. 36

Honeysuckle 146

St. John's-wort 88

Azalea 104

Buffalo-berry 95

Opposite Leaves without Teeth

Snowberry 145

Honeysuckle 146

Honeysuckle 146

Honeysuckle 146

leaves alternate,
but clustered at ends of twigs;
often hairy along edges

Azalea 104

Honeysuckle 146

Privet 132

St. John's-wort 88

St. John's-wort 88

leaves alternate, but often clustered at ends of twigs

underside of leaves densely woolly; wool green at first, then rusty orange; edges of leaves curl under

leaves alternate, but often in clusters

St. Peter's-wort 89A

St. Peter's-wort 89A

Azalea 104

Labrador Tea 116A

green above, whitish beneath

Laurel 110

Honeysuckle 146

mostly alternate

Coralberry 145

Mistletoe 17A

Sand-myrtle 113B

Key #3—**FLOWERS**

This Key is arranged as follows:

    I   Flowers mostly in small clusters along the stems, pp. 1-3

   II   Flowers in the form of Catkins, pp. 4-5

  III   Flat flower clusters mostly at the ends of the stems, pp. 6-7

  IV   Rounded or elongated flower clusters mostly at the ends of the stems, pp. 8-12

   V   Widely open flowers with distinct petals, pp. 13-17

  VI   Bell-shaped flowers, pp. 18-21

 VII   Trumpet-shaped or deeply cut flowers, pp. 22-23

VIII   Pea-type flowers, p. 24

Note: Within each of the above groups the earliest blooming flowers are placed first, the latest blooming ones last. Flower colors are noted when distinctive.

The divisions and arrangement of this Key were determined solely to make finding any particular flower easier; the headings were chosen for the same purpose and are not intended to be technically descriptive nor scientifically correct. The botanical nomenclature for floral parts is extensive, precise, and extremely important for the taxonomist but hardly practical for anyone unused to the subject. It is well, however, to know that shrubs produce three main types of flowers:

    (1) Perfect: both male and female parts in the same flower

    (2) Male (staminate)

    (3) Female (pistillate)

Some shrubs bear only perfect flowers; others are *monoecious,* producing both male and female flowers on the same plant. Some are *dioecious,* bearing male flowers on one plant and female ones on a separate plant. The dioecious shrubs, of course, bear fruit only on the female plants; the other two types can produce fruit on all fertile plants. There are occasional plants which have combinations of the above possibilities, but usually these require two or more plants to produce fruit and so for practical purposes are essentially dioecious. It is also true that, regardless of the type of flowers, fertilization is best achieved, and therefore plants are more productive of fruit, when several plants of the same species are present. Although the blooming period of most plants is brief, all mature shrubs should produce flowers, whereas not all bear fruit.

I. FLOWERS MOSTLY IN SMALL CLUSTERS ALONG THE STEMS

greenish yellow

pale yellow

inconspicuous, pale yellow

yellow

This is the first shrub to bloom (late winter and early spring).

fragrant; pinkish purple

Spice-bush 21

Buffalo-berry 95

Prickly Ash 70

Daphne 91B

Barberry 18

male (staminate)

Buckthorn 84

female (pistillate)

Mountain Holly 83

Oleaster 96

Euonymus 80

FL
2

Holly 78

Euonymus 80

deep
pink

Laurel 110

Holly 78

Buckthorn 84

male
(staminate)

female
(pistillate)

Holly 78

Flowers in Small Clusters along the Stems

Both male and female flowers are produced in each individual cluster.

male (staminate)

female (pistillate)

pinkish purple

Coralberry 145

Groundsel-bush 155

Marsh-elder 156

yellow

These are the last two shrubs to bloom (late in the fall).

rose purple

Witch-Hazel 38

Swamp Loose-strife 97B

Mistletoe 17A

II. FLOWERS IN THE FORM
   OF CATKINS

Note:
Alders may
be found
blooming
from March
to Sept.

female
(pistillate)

male
(staminate)

female
(pistillate)

male
(staminate)

female
(pistillate)

male
(staminate)

Alder 6

Hazelnut 10

Sweet Gale 4

female
(pistillate)

male
(staminate)

This row is all Willow 1.

FL
4

male (staminate)

female (pistillate)

female (pistillate)

Birch 9

male (staminate)

Sumac 74

male (staminate)

Sweet Fern 11B

female (pistillate)

Hornbeam 12

Oak 14

female (pistillate)

male (staminate)

female (pistillate)

Bayberry 4
(also Wax-myrtle)

female (pistillate)

male (staminate)

Alder 6

Chestnut 13
(Chinquapin)

male (staminate)

FL
5

III. FLAT CLUSTERS OF FLOWERS,
Mostly at the Ends of the Stems

Viburnum 136

Firethorn 42

Note sterile marginal flowers.

Sand-myrtle 113B

Viburnum 136

Note:
Flowers of all Viburnums have five petals.

Note:
Flowers of all Dogwoods have four petals.

Viburnum 136

Dogwood 98

Elderberry 134

FL
7

Hop-tree 71

Note
sterile marginal
flowers.

Ninebark 27

Spirea 40

Hydrangea 28

FL 8

brownish
purple

Lilac 130

Shrub Yellow-root 25

Elderberry 134

Rounded or Elongated Flower Clusters
at Ends of Stems

Cherry 54

Currant 30

yellow

Currant 30

Buffalo-nut 16

Barberry 18

Note tiny
pink or
brownish
hairs.

Fringe-tree 129

Smoke-tree 72

FL
9

75

purple with golden pollen

FL 10

Virginia-willow 23B

False Indigo 66

Dogwood 98

Spirea 40

New Jersey Tea 45A

Privet 132

Deutzia 26

Rounded or Elongated Flower Clusters at Ends of Stems

Beware of this shrub; it is worse than Poison Ivy for many people.

Buttonbush 144

Privet 132

Poison Sumac 74

buds covered with sticky red hairs

pink

False Spirea 24

Bramble 46

Spirea 40

Clethra 102

FL 11

Knotweed 17B

½ actual size

Hydrangea 28

Marsh-elder 156

Sumac 74

Sumac 74

Sarsaparilla 92

yellow

Forsythia 131

Shadbush 44

Cherry and Plum 54

Jetbead 45B

Chokeberry 39

Dogwood 98

Hawthorn 43

FL
14

Euonymus 80

Labrador Tea 116A

pink

purple, quickly turning to tan

Matrimony-vine 125B

Euonymus 80

Laurel 110

Bramble 46
(Raspberry)

FL
15

Bramble 46
(Blackberry)

Laurel 110

Mock-orange 22

Euonymus 80

Widely Open Flowers

magenta

Most wild roses are pink or white.

Rose 57

Rose 57

Rose 57

buds and flower stems covered with sticky red hairs

magenta, fading to lavender pink

Bramble 46 (Raspberry)

Bramble 46 (Raspberry)

yellow

yellow;
five petals

yellow;
four petals

(smaller and
narrower in
some species)

Cinquefoil 64

St. John's-wort 88

St. Peter's-wort 89A

double and single
forms

Color varies
from white to
pink, rose, blue
or purple.

Rose of Sharon 90

FL
18

Leatherleaf 116B

Andromeda 114

yellow

Leatherwood 91A

Andromeda 114

Huckleberry 126

Silverbell 128

Blueberry 119

Bladdernut 87

Blueberry 119

Blueberry 119
(Deerberry)

Bell-Shaped Flowers

red (to pink)

Leucothoë 118   Huckleberry 126

Gooseberry 30   Gooseberry 30

greenish
yellow, tipped
with red or
pink

Currant 30

Menziesia 103

Leucothoë 118

Currant 30

Bog Rosemary 113A

Huckleberry 126

Huckleberry 126

Lyonia 117

pink

pinkish
purple

Snowberry 145

Blueberry 119
(Bilberry)

Lyonia 117

Coralberry 145

FL
21

pale to
deep rose
purple

pale
yellow

Azalea 104
(Rhodora)

Honeysuckle 146

Note: Azaleas
vary from
white to deep
pink, yellow or
orange red.

Azalea 104

white
to pink

red brown

Sweet Shrub 20

Honeysuckle 146

pale yellow,
turning deep yellow,
orange or red

Note: Leaves
have
teeth.

Note: Leaves of true
Honeysuckles have
no teeth (entire).

Bush-honeysuckle 154

yellowish
white

Honeysuckle 146

pale
yellow

Honeysuckle 146

Honeysuckle 146

purple or
magenta

white, pink
or purple

Rhododendron 104

Rhododendron 104

VIII. PEA-TYPE FLOWERS

yellow

Gorse 67B

yellow

yellow

Scotch Broom 65B

reddish pink
to pale pink

Redbud 68

pink

Bristly Locust 69

yellow

Dyer's Greenweed 65A

Key #4—FRUIT

This Key is arranged as follows:

  I Acorns, Nuts, p. 1

 II Pods, Capsules, and Similar Fruit, pp. 2-8

III Berry-like Fruit, pp. 9-20

IV Miscellaneous Fruit, pp. 21-23

Note: The first and fourth groups are short; the second and third are longer. A little study, however, will indicate that the individual fruits in both the capsule and berry-like sections are distinctive, and few, if any, are so alike as to cause confusion. Note in both these sections the type of growth of the whole cluster, whether branching from one point or from several points, or not branching and arranged along a single axis. Note shape and size of the individual capsules and berries. The word "bloom" is often used to describe a whitish, powdery covering often found on berries (and other parts of a plant). This bloom, when present, can be rubbed off revealing the true color beneath.

The fruit of many plants persists for a considerable part of the year and therefore is a valuable aid to identification; even a dried berry may be found in winter, which at least eliminates other types of fruit for that particular plant.

I. ACORNS, NUTS

Oak 14

Oak 14

Chestnut 13
(Chinquapin)

poisonous

Buffalo-nut 16

Hazelnut 10

Hazelnut 10

Redbud 68

Bristly Locust 69

Gorse 67B

Dyer's Greenweed 65A

Scotch Broom 65B

FR
2

Sweet Shrub 20

Bladdernut 87

Silverbell 128

False
Indigo 66

Witch-Hazel 38

Forsythia 131

Lilac 130

Rose of Sharon 90

Rhododendron 104

Mock-orange 22

Azalea 104

FR
3

St. John's-wort 88

St. John's-wort 88

Menziesia 103

St. Peter's-wort 89A

St. Peter's-wort 89A

Sand-myrtle 113B

Leucothoë 118

Note the odd way these capsules open, hinging from the outer end.

Labrador Tea 116A

Bush-honeysuckle 154

False Indigo 66

Virginia-willow 23B

FR 4

FR
5

Laurel 110

Laurel 110

Andromeda 114

Leatherleaf 116B

Laurel 110

Andromeda 114

Bog Rosemary 113A

Cinquefoil 64

Leucothoë 118

Lyonia 117

Ninebark 27

red at first,
then greenish,
finally brown ←

Lyonia 117

FR
6

Mock-orange 22   Swamp Loose-strife 97B   Marsh-elder 156   Deutzia 26   Clethra 102

FR 7

Shrub Yellow-root 25

False Indigo 66

Spirea 40

False Spirea 24

Note: This almost always looks messy, full of cobwebs and other foreign matter.

False Indigo 66

remnants of
sterile flowers

few actual
fruit

FR
8

Spirea 40

Hydrangea 28

Spirea 40

Hydrangea 28

red

red, rarely
yellow

Plum 54

FR
9

red, yellow
or green

Gooseberry 30

Rose 57

Plum 54

red
to purple

red

dark
blue

Viburnum 136

Fringe-tree 129

yellow to pink,
finally dark
blue with bloom

red

Viburnum 136

blue

Rose 57

FR
10

Dogwood 98

orange red

white

red, turning
dark purple or black

Firethorn 42

Dogwood 98

Viburnum 136

Berry-like Fruit

black

dark
blue

Viburnum 136

Elderberry 134

three types:
red, purple
or black

dark blue
or black,
sometimes
with bloom

Chokeberry 39

Viburnum 136

shiny
black

pink, turning
dark blue with
heavy bloom

Viburnum 136

Viburnum 136

red

shiny black

black with bloom, giving bluish cast

Privet 132

Privet 132

Privet 132

FR 12

blue black on red stems

Elderberry 134

Dogwood 98

red

white

pink, turning dark blue with bloom

Dogwood 98

Rose 57

Viburnum 136

103

red, turning
blue black

Cherry 54

dull
white

Poison Sumac 74

pink or red,
turning
purple
or black

red

Barberry 18

Shadbush 44

black

Currant 30

red

Currant 30

red, yellow
or green

red

red; sometimes
bristly-hairy

Hawthorn 43

Rose 57

Rose 57

red

red;
bristly

red; sometimes
bristly-hairy

Rose 57

Currant 30

Rose 57

black;
bristly

dull red;
prickly

Currant 30

Gooseberry 30

Berry-like Fruit

red

red

red

Mountain Holly 83

Honeysuckle 146

Honeysuckle 146

All Blueberries
contain many
small seeds.

All Huckleberries
are ten-seeded.

blue with bloom

← greenish (or slightly
purplish) with bloom

Blueberry 119
(Deerberry)

Huckleberry 126

dull red

dull
red

blue
with
bloom

Honeysuckle 146

Gooseberry 30

Gooseberry 30

FR
15

black

All Buckthorn berries contain two to four seeds.

All Blueberries contain many small seeds.

All Huckleberries are ten-seeded (see previous page).

blue with bloom

**Blueberry 119
(typical Lowbush)**

shiny black

Huckleberry 126

two types: blue with bloom or shiny black

Buckthorn 84

red

Honeysuckle 146

**Blueberry 119
(Highbush)**

black

red, dotted with silver or brown scales

yellow or silvery

red or yellow

Buckthorn 84

Oleaster 96

Oleaster 96

Buffalo-berry 95

Berry-like Fruit

blue
with
bloom

All Blueberries
contain many
small seeds.

All Huckleberries
are ten-seeded.

red

**Blueberry 119
(Bilberry)**

blue
with
bloom

black

Huckleberry 126

Honeysuckle 146

black

Huckleberry 126

Holly 78

red

red

red

red

Holly 78

Daphne 91B

Honeysuckle 146

Holly 78

red;
fuzzy

Sumac 74

maroon or
pinkish purple

Coralberry 145

black; almost
always four
berries together

Jetbead 45B

white

Snowberry 145

dull white

Poison Sumac 74

grayish;
waxy

Bayberry 4
(also Wax-myrtle)

white;
sticky
when
crushed

Mistletoe 17A

white

Dogwood 98

FR
18

Berry-like Fruit

purple to
black

Cherry 54

yellow to
red

red

Matrimony-vine 125B

red

silvery

Barberry 18

Oleaster 96

green, yellowish
or red

Spice-bush 21

red

red

Leatherwood 91A

Barberry 18

Dogwood 98

red

black

Honeysuckle 146

blue
with
bloom

red

Viburnum 136

Honeysuckle 146

Euonymus 80

FR
19

seeds with
orange skin

Berry-like Fruit

black

reddish
pink

Euonymus 80

pink

Euonymus 80

black
or red

red

FR
20

black
when ripe

Bramble 46
(Raspberry)

Bramble 46
(Raspberry)

½ actual
size

black

Bramble 46
(Blackberry)

red

Bramble 46
(Raspberry)

Sarsaparilla 92

½ actual
size

Sarsaparilla 92

IV. MISCELLANEOUS FRUIT

Seed heads produce cotton (pappus) which is carried by the wind, distributing the seed.

winged seed

Hornbeam 12

winged seed

Knotweed 17B

Groundsel-bush 155

willow fruit in various stages of development

cotton

Willow 1

Small section of a large fruiting plume. Each of the many small branches is covered with fine hairs, giving the whole plume a "smoky" appearance.

few actual seeds

red

Smoke-tree 72

Sumac 74

Sumac 74

FR 22

black case

three whitish nutlets

red case turning brown and opening to expose shiny black seeds

persistent silvery-lined base

winged seed

Hop-tree 71

New Jersey Tea 45A

Prickly Ash 70

winter catkins that produce flowers later in the year

female (pistillate)

male (staminate)

fruiting cones

actual fruiting cones

produced from the female flowers of the previous year

Alder 6

Note: This is not a fruiting cone, but a gall formed around insect larvae (see actual Willow fruit, p. 21).

FR 23

Sweet Fern 11B

Sweet Gale 4

orange skin over seed

reddish-pink case

Buttonbush 144

Willow 1 (see note above)

Euonymus 80

orange skin over actual seed

pink

Birch 9

Euonymus 80

114

Key #5—TWIGS

(Evergreens not included)

This Key is divided into two main sections:

I ALTERNATE: buds (and leaf-scars) arranged alternately along the twigs, pp. 1-15

Note: Pages 1-4 include twigs which are remarkable for one reason or another. Beginning on page 5 they are arranged in order of size of twigs and buds, ending with twigs which die back in winter, or whose buds are inconspicuous or entirely invisible, being concealed within the twig itself.

II OPPOSITE: buds (and leaf-scars) opposite, or in whorls of three or four around the twig at each node (growth point), pp. 16-21

Leaf-scars are left at the point of separation when the leaves drop off in the fall. They vary greatly in size and shape and are often typical of a particular plant.

Buds develop during the summer at the ends of twigs and also along the twigs, usually just above the junction of the leaf-stalk and twig (in the axil thus formed), and therefore are found just above the leaf-scar on the winter twig. A few plants produce buds inside the leaf-stalk itself and thus appear in the middle of the leaf-scar.

Buds, to the untrained eye, may appear much alike, but a little scrutiny reveals that each plant has its own particular type of bud, and therefore buds are an extremely important aid to identification. A lens shows these details clearly, but even without one many differences are apparent. Buds contain the whole potential of growth, leaves and flowers, and it is therefore no wonder that they are distinctive.

Buds from which flowers will emerge are often markedly different (usually larger) than leaf (or growth) buds, and when present are very characteristic of specific plants.

Buds at the ends of the twigs are often larger and more distinctive than the side (axillary) buds.

Buds are usually protected in winter by an outer layer of overlapping scales, the number, shape, arrangement and color of which are often characteristic. A few plants have "naked" buds (without scales), and among the shrubs included in this book the Willows are the only plants with buds covered by only a single scale.

Twigs (and buds) vary considerably in color, but for our purpose, unless note is made of color, it can be assumed that the twigs are some type of brown or brownish gray. Throughout this Key note is often made of red (maroon) *and* green twigs. The reason for this apparently startling difference of color on the same twig results from the fact that some twigs which are normally green in the shade turn red in the sun (or vice versa), and frequently one half of a twig will thus be red, the other (the side not exposed to the sun) will be green. As with other parts of a plant, twigs are often covered with a whitish bloom which can be rubbed off, revealing the true color beneath.

The word twig, as used here, applies largely to the latest growth of the shrub. This is easily determined, as the new growth of almost all shrubs is different in color and texture from that of the older wood, and the junction between the old and the new growth is almost always marked by a ring around the twig at that point. It should also be noted that almost all shrubs produce side branches only from two-year wood (or older). Two remarkable exceptions to this are seen on page 2. Both Alternate-leaved Dogwood and Clethra branch the first year and this in itself marks these two plants apart from other shrubs.

I. ALTERNATE SECTION

long, pointed buds; side buds
close to stem (appressed)

brown buds;
twigs smooth

red buds;
twigs smooth

red buds;
twigs fuzzy

The buds of all Willows have
only one scale. (They are the
only plants in this book with
this characteristic.)

Shadbush 44 **Chokeberry 39** **Chokeberry 39** **Willow 1**

These two plants have naked buds (buds without any
scales). (They are the only shrubs shown here with alternate
buds that have this characteristic.)

dark brown buds mostly
opposite, but occasionally
partly alternate (see also
Fringe-tree, p. 5)

Witch-Hazel 38 **Buckthorn 84** **Buckthorn 84**

TW
1

116

Alternate Section

These two plants produce side branches on the new growth. (They are the only plants in this book that normally do this.)

Matrimony-vine 125B

Dogwood 98
(Alternate-leaved)

Clethra 102

These two plants usually have spurs on which flowers and fruit are produced.

Bayberry 4

Holly 78

Mountain Holly 83

Note: The plants on this page commonly produce buds in clusters at the ends of the twigs.

flower buds

leaf buds

buds covered with a bloom, giving grayish or pinkish look; scales indistinct

hairy twigs

Azalea 104

Azalea 104 (Rhodora)

Menziesia 103

Oak 14

TW 3

usually several side buds from one point (node)

aromatic twigs

aromatic twigs

flower buds

leaf buds

Cherry 54

Plum 54

Currant 30

Bayberry 4

Spice-bush 21

Birch 9

Note: The plants on this page commonly produce catkins, which are evident in winter. These develop later in the season as male (staminate) flowers. Some Alders also have catkins that develop into female (pistillate) flowers.

male (staminate)

female (pistillate)

slightly undeveloped

Alder 6

Alder 6

Hazelnut 10

strongly aromatic, reddish-brown twigs

slightly aromatic, dark gray twigs

pungently aromatic twigs

Sweet Fern 11B

Sweet Gale 4

Birch 9

Sumac 74

Hazelnut 10

TW 4

Alternate Section

Note bud in *middle* of leaf-scar.

Note bud *above* *large* leaf-scar.

Note bud *above* *small* leaf-scar.

mostly opposite, but occasionally partly alternate

Sarsaparilla 92 **Sumac 74** **Sumac 74** **Poison Sumac 74** **Sumac 74** **Smoke-tree 72** **Fringe-tree 129**

leaf bud

flower bud

Note: All Willows have a single scale covering the bud. This comes off as a cap, or splits to allow the expanding growth to emerge. The "pussy" is the early stage of the Willow flower (see p. 4, Flower Key).

Willow 1

Buffalo-nut 16

False Spirea 24

Alder 6

TW 5

120

disagreeable odor
when broken

bark of twigs and
stems peeling

Currant 30   Cherry 54   Currant 30   Alder 6

inner bark of
twigs and stems
very yellow

red buds;
skunk odor when
broken

usually brown,
sometimes pink
buds

dark red
buds

Currant 30   Shrub Yellow-root 25   Currant 30   Shadbush 44   Chokeberry 39

Daphne 91B   Alder 6   Silverbell 128   Alder 6

TW 7

twigs usually
hairy

tips of buds
whitish

red buds

Hazelnut 10   Alder 6   Hazelnut 10   Lyonia 117

buds usually
clustered at tips
of twigs

red buds;
red and green twigs

one scale
covers each bud

Oak 14   Chestnut 13   Rose 57   Willow 1   Buckthorn 84

twigs somewhat
hairy (best seen
with lens)

often several buds
at each node (joint)

flower
buds

Hornbeam 12   Birch 9   Plum 54   Plum 54   Leucothoë 118

TW
8

Alternate Section

Note: Blueberry twigs are red and green. Two-year-old stems also are often colored: red, green, yellow and brown. This variety of coloration of the older wood is typical of Blueberries. Huckleberries, which are often confused with Blueberries, do not have this characteristic (see note below). (See Bark Key, pp. 2 and 18.)

flower buds   leaf buds

Blueberry 119

flower buds   leaf buds

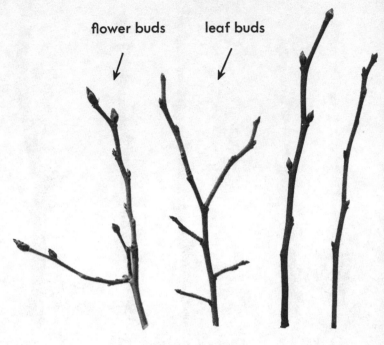

Blueberry 119

very dark twigs, often with silvery coating

red buds; twigs and older stems (light) brown

Note: Huckleberry twigs are typically red (sometimes pinkish, brown or dark gray). All other growth (two years and older) is uniformly dark gray (or brown) (see Blueberry note above). (See Bark Key, p. 16.)

leaf buds

flower buds

Holly 78   Lyonia 117   Huckleberry 126

buds and twigs
golden brown
or silvery

red buds;
twigs (light)
brown

buds and twigs
pinkish with bloom

buds and twigs
pinkish with bloom

one scale
covers each bud

Oleaster 96   Lyonia 117   Huckleberry 126   Blueberry 119
(Deerberry)   Willow 1

TW 10

often two buds
at each node
(joint), a
larger one
above a very
small one;
aromatic twigs

conspicuously
glandular-
dotted twigs

reddish-brown twigs;
strongly aromatic

dark gray twigs;
slightly aromatic

red buds;
red and green twigs

Birch 9   Sweet Fern 11B   Sweet Gale 4   Rose 57   Buckthorn 84   Spice-bush 21

silvery,
fuzzy twigs

Blueberry 119
(Bilberry)

Blueberry 119
(Bilberry)

Oleaster 96

red and green twigs
covered with many
minute dots

See note p. 9
concerning
Blueberries and
Huckleberries.

reddish twigs without
(or with slight) bloom

flower buds

Blueberry 119

Huckleberry 126

Leucothoë 118

Note rounded or relatively
blunt end buds.

Note sharp-pointed
end buds.

pungently
aromatic

Holly 78   Holly 78   Mountain Holly 83   Sumac 74

TW
12

Note: Large flower buds are on
larger stems; only very small
leaf buds are produced on twigs.

Redbud 68   Hop-tree 71   Leatherwood 91A

fruit remnants

twigs fuzzy;
buds shaggy

twigs red and green

buds round
and red; at
tips of twigs

TW
13

The leaf-stalk often
persists after the
rest of the leaf
has fallen.

reddish and
green twigs

green twigs

fruit
remnant
saucer-
shaped;
silvery
inside

green, angled
(several-sided)
twigs

Cinquefoil 64

Bramble 46
(Blackberry)

Hawthorn 43

Virginia-willow 23B

Groundsel-bush 155

New Jersey Tea 45A

Scotch Broom 65B

fruit remnants

Matrimony-vine 125B

False Indigo 66

False Indigo 66

green, angled (several-sided) twigs

twigs silvery, fuzzy

Dyer's Greenweed 65A

Oleaster 96

Rose of Sharon 90

The twigs shown on this page
largely die back in winter, but buds
can usually be found on larger stems.

False Spirea 24

Ninebark 27

**Bramble 46
(Raspberry)**

TW
15

mostly
opposite,
but upper
twigs with
alternate
characteristics

Marsh-elder 156

Spirea 40

Elderberry 134   Lilac 130   Viburnum 136   Bladdernut 87

naked buds
(buds without scales)

maroon or
green twigs

flower bud   leaf bud

flower
bud

leaf
bud

Viburnum 136   Dogwood 98   Viburnum 136   Viburnum 136

Opposite Section

Note: Side buds in the top row are markedly appressed (close to stem).

flower bud

growth (leaf) buds

pinkish or
yellow twigs

bright red twigs
(yellow in variety)

Viburnum 136　　　　Dogwood 98　　　　Viburnum 136　　　　Dogwood 98

Forsythia 131　　　　Viburnum 136　　　　Honeysuckle 146　　　　Elderberry 134

opposite, but
often partly
alternate

red, maroon or
green twigs

Buckthorn 84

Viburnum 136

Dogwood 98

Viburnum 136

opposite, but sometimes
partly alternate

this type
of growth
sometimes,
but not
always,
found

Hydrangea 28

Fringe-tree 129

Honeysuckle 146

fruit (or remnants)
present most of year

twigs and buds
usually silvery

green or maroon twigs;
often four-sided

Bush-honeysuckle 154   **Buffalo-berry 95**   **Honeysuckle 146**   **Dogwood 98**

Jetbead 45B   **Honeysuckle 146**   **Euonymus 80**   **Privet 132**

Opposite Section

twigs with slight ridge
(or small wing) along
opposite sides

Honeysuckle 146 Honeysuckle 146 Snowberry 145 Coralberry 145 St. John's-wort 88

green or maroon
twigs

fuzzy twigs

Euonymus 80 Deutzia 26 Privet 132 Privet 132 Hydrangea 28

Opposite Section

buds often in whorls of three or four around stems at each node (joint)

aromatic twigs

maroon and green twigs

Buttonbush 144

Sweet Shrub 20

Mock-orange 22

Dogwood 98

distinctive wings alternating at right angles along twigs

large, distinctive wings
(see also Bark Key, p. 23)

mostly opposite, but upper twigs with alternate characteristics

St. Peter's-wort 89A

Euonymus 80

Marsh-elder 156

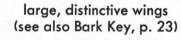

Key #6—**BARK**

This Key is divided into two main sections (determinable at any time of the year):

  I ALTERNATE: barks of shrubs having alternate characteristics (leaves, buds and leaf-scars), pp. 1-18

II OPPOSITE: barks of shrubs having opposite characteristics, pp. 19-27

    Within the above sections, similar barks are grouped together, such as shaggy, striped, spotted, nondescript, or relatively smooth barks. It will be apparent from a perusal of this Key and an examination of actual shrubs that no exact grouping is possible. Bark is not a detail which is as definite and positively characteristic as the details shown in other Keys, although many barks are truly distinctive. The Bark Key should be used last to verify identification made in previous Keys and, used in this way, will prove invaluable, often being the deciding factor in questions of doubt. Used first, it will only prove confusing. However, unlike the other details, barks are always present.

    Two (or more) barks of each plant are always shown together. The bark of larger stems of some shrubs is quite different from that of the smaller (or younger) growth, while on other shrubs the bark is consistently uniform throughout. Therefore always check both sizes, the larger one first, to determine the position in the Key. In selecting the barks for this Key a definite size limit was established for the larger barks. After a certain size, most barks become very roughened and are less distinctive than on the smaller and younger stems. Therefore, in comparing barks of actual shrubs with the pictures shown here, keep within the sizes indicated. Actually, stems of the correct size will prove to be much more accessible than the larger ones, concealed in the middle of a shrub which is often almost impenetrable.

    Some shrubs commonly have two or more quite different kinds of bark on different plants of the same species. There are a number of reasons for this. Sometimes it is a question of location or environment, but these differences may also be seen on plants apparently growing under similar conditions. In such cases, a variety may be indicated or possibly a particular species is not a stable one and is still undergoing evolutionary changes. Occasionally, the bark of the male (staminate) plant is decidedly different from that of the female (pistillate) plant. In any event, regardless of cause, if a consistent variation is commonly encountered, both types of bark will be found in appropriate places in this Key. (*Note:* No attempt has been made to explain the individual differences shown here, as this is a little-explored subject outside the scope of this book. But it would be an interesting one to pursue. It might even lead to important discoveries concerning the nature of plants and their evolution.)

    Look at both sides of the stem, as bark in the sun is often different from that on the shaded side. One last word of caution: Do not try to guess what a bark covered by lichens, fungus or moss really looks like; find a section not covered over, or remove the growth carefully.

I. ALTERNATE SECTION

The plants on these two pages are all closely related. Their barks, although not identical, have much in common. It is well to become familiar with this type of bark, as it distinguishes these plants from others in this section.

Laurel 110

Andromeda 114

Rhododendron 104

Rhododendron 104

green, red
or maroon

Alternate Section

pale
brown

Blueberry 119

Clethra 102

Lyonia 117

Lyonia 117

Azalea 104

BK
2

139

Ninebark 27

Cinquefoil 64

Oleaster 96

Oak 14

small stems
green or maroon

**Bramble 46
(Raspberry)**

Sand-myrtle 113B

Leucothoë 118

Smoke-tree 72

Hawthorn 43

BK
4

green

Matrimony-vine 125B

Groundsel-bush 155

Scotch Broom 65B

BK
5

Chokeberry 39

Plum 54

Cherry 54

Silverbell 128

Shadbush 44

Currant 30

Dogwood 98
(Alternate-leaved)

Rose 57

BK
6

Alternate Section

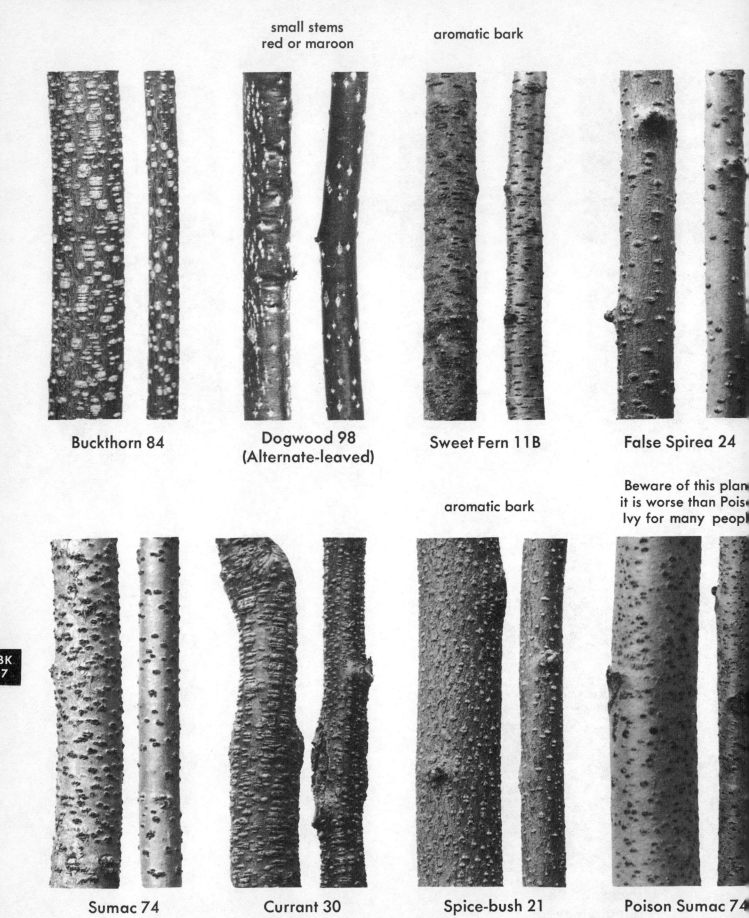

small stems
red or maroon

aromatic bark

Buckthorn 84

Dogwood 98
(Alternate-leaved)

Sweet Fern 11B

False Spirea 24

Beware of this plan
it is worse than Pois
Ivy for many peopl

aromatic bark

Sumac 74

Currant 30

Spice-bush 21

Poison Sumac 74

BK
7

aromatic bark

Bayberry 4   Bristly Locust 69   Firethorn 42

mostly opposite, but
often partly alternate

Fringe-tree 129   Sumac 74   Birch 9   Alder 6

BK
8

145

lenticels (dots on bark)
usually orange

large, soft
pith

small stems
often hairy

Alder 6

Holly 78

Sumac 74

large, soft pith

large, soft pith

BK
9

Sumac 74

Sumac 74

Hop-tree 71

aromatic bark

Mountain Holly 83   Holly 78   Sweet Gale 4

aromatic bark

Oak 14   Oleaster 96   Wax-myrtle 4

BK
10

Alternate Section

Chokeberry 39

Oleaster 96

Buckthorn 84

Plum 54

Buffalo-nut 16

Sarsaparilla 92

BK
11

aromatic bark

Willow 1　　　　　Spice-bush 21　　　　　Alder 6

Holly 78　　　　Chestnut 13
　　　　　　　　(Chinquapin)　　　　　Oak 14

BK
12

Alternate Section

red-brown bark

Currant and Gooseberry 30

Hazelnut 10

Hazelnut 10

red-brown bark

Buckthorn 84

Hornbeam 12

Witch-Hazel 38

BK 13

Beware of this plant; it is worse than Poison Ivy for many people. It is found almost entirely in swamps and wet ground.

pungently aromatic bark

Poison Sumac 74

Smoke-tree 72

Fragrant Sumac 74

bark very strong and pliable; wood soft and weak

small stems green, tinged with red

Leatherwood 91A

Virginia-willow 23B

Rose of Sharon 90

BK
14

151

inner bark yellow

Daphne 91B

Shrub Yellow-root 25

Azalea 104

Note flower
buds on
large stems.

small
stems
green

BK
15

Redbud 68

New Jersey Tea 45A

Dyer's Greenweed 65A

Buckthorn 84   False Indigo 66   Labrador Tea 116A   Leatherleaf 116B

False Indigo 66   Willow 1   Leucothoë 118   Huckleberry 126

BK
16

Alder 6

Clethra 102

Azalea 104

stems rust-colored; flaky bark

stems reddish
brown; flaky bark

**BK
17**

Hornbeam 12

Menziesia 103

Currant 30

stems bamboo-like;
hollow

stems red, maroon or
green; solid pith

stems golden brown or
red (sometimes with
bloom)

stems reddish brown;
skunk odor when
bruised

Knotweed 17B

Bramble 46
(Blackberry)

Bramble 46
(Raspberry)

Currant 30

small stems green, red,
maroon, yellow or brown

old stems mostly
light gray

young growth brown
or golden brown

Blueberry 119

Spirea 40

BK
18

II. OPPOSITE SECTION

Deutzia 26

St. John's-wort 88

St. Peter's-wort 89A

Buttonbush 144

Coralberry 145

Hydrangea 28

Honeysuckle 146

Bush-honeysuckle 154

Mock-orange 22

mostly alternate, but occasionally partly opposite

actually alternate, but leaves always clustered at ends of stems, giving a definite opposite appearance

Laurel 110

Andromeda 114

Rhododendron 104

BK
20
Opp

Snowberry 145   **Hydrangea 28**   **Viburnum 136**

**BK
21
Opp**

Lilac 130   **Buffalo-berry 95**

Dogwood 98

Dogwood 98

Viburnum 136

Viburnum 136

BK
22
Opp

vigorous young shoots green
with four light tan stripes

Euonymus 80

Honeysuckle 146

Euonymus 80

stems (small ones especially)
greenish with dull orange stripes

small stems red, maroon or green

Bladdernut 87

Euonymus 80

Dogwood 98

BK
23
Opp

large pith

Elderberry 134

Viburnum 136

Buckthorn 84

large pith

Buffalo-berry 95

Viburnum 136

Elderberry 134

BK
24
Opp

aromatic bark

Jetbead 45B

Fringe-tree 129

Sweet Shrub 20

large and small stems red

large and small stems
pinkish or greenish yellow

Forsythia 131

Dogwood 98

Dogwood 98

BK
25

Opp

Privet 132

Viburnum 136

Euonymus 80

Privet 132

Viburnum 136

Privet 132

BK
26
Opp

Viburnum 136   Dogwood 98

green

Viburnum 136   Marsh-elder 156   Mistletoe 17A

PART II

MASTER PAGES

(all actual size except when noted)

BROAD-LEAVED UPRIGHT SHRUBS, pp. 1-156

NEEDLE-LEAVED SHRUBS (Upright and Prostrate), pp. 157-162

BROAD-LEAVED GROUND COVERS, pp. 163-171

VINES (all Broad-leaved), pp. 172-200

MASTER PAGES

The Master Pages have two main uses: (1) to assemble in one place the important features of each shrub, and (2) to identify species.

At the heading of each Master Page the common name is given, followed by the botanical one. This is followed by a flower note: whether *perfect* (both male and female parts in the same flower), *monoecious* (distinct male and female flowers on the same shrub), or *dioecious* (male flowers on one plant, female flowers on a different one). The approximate date of blooming is given. As the dates vary somewhat in different years (especially for the early-bloomers) and as the same shrubs bloom at different times in different parts of their range, no exact date can be given. Comparisons with a few known dates of bloom will indicate more exactly what to expect in any particular section of the country.

Note is made as to whether the shrubs are alternate or opposite. A description of range is given, but it should be realized that the outside limits indicated should not be taken to mean that any particular shrub will be found throughout the area. Some plants grow only at high elevations within their indicated range, others only at low altitudes; some only in wet soil, others only in dry soil, etc.

The heights given for each shrub are for average mature plants, and, of course, occasional giants can be found of almost any plant. However, a plant indicated as 2-3 feet will rarely be found 5 feet tall, and the plants that are normally very large when full grown will certainly show evidence of this at an early age.

The headings are followed by the pictures of actual details, and in cases where several species are included in one genus, comparison of these details leads to species identification. As explained before, it is by comparing all available details that final identification is achieved, and the combination of determining factors is not the same in each case, depending on the season or the characteristics of the plant itself. Throughout the Master Pages, notes point out differences among species, when identification might be difficult otherwise.

WILLOW — *Salix* Dioecious   Alternate (rarely sub-opposite); few foreign species

Beak Willow *Salix bebbiana*	late April-early June	Nfd. to Alaska, s. to Md., Ill., Ia. & west 25 ft.
Bearberry Willow *S. uva-ursi*	June-July	Lab. to Alaska, s. to alpine summits of N.H. & N.Y. Prostrate
Dwarf Gray Willow *S. tristis* (Dwarf Upland W.; Dwarf Pussy W.) (Considered by some botanists as a variety of Gray Willow.)	March-April	Me. to Minn., s. to Fla. & Okla. 1½ ft.
Gray Willow *S. humilis* (Upland W.; Prairie W.)	March-early June	Nfd. to Minn., s. to N.C. & Kans. 8 ft.
Pussy Willow *S. discolor* (Glaucous W.)	late Feb.-May	N.S. to Man., s. to Va., Mo. & west 20-25 ft.
Shining Willow *S. lucida*	late April-June	Lab. to Man., s. to N.J. & Ia. 20 ft.
Silky Willow *S. sericea*	late March-early May	Que. to Mich., s. to S.C. & Mo. 12 (occ. 15) ft.

There are a great many named species of Willows, Rehder's manual listing sixty-three, and Gray's manual fifty-four. Both manuals list many more hybrids and forms, so that the total number is truly impressive. However, as explained in the Introduction to this book, certain genera contain plants that do not consistently remain true to any species type. The Willows are such a genus, and therefore only a few species are given here to indicate typical forms and the means by which these species, if found true to type, can be identified.

The method used here for the Willows is based primarily on summer characteristics. Winter identification of species is difficult and should not be attempted at first. It is interesting to note, however, that the detail most helpful in distinguishing any Willow from other plants is often considered a winter characteristic: the single bud scale that covers the individual buds of all Willows. All other shrubs included in this book have two or more scales on each bud, or have naked buds (buds without scales). Although buds are more evident in winter, they can usually be found on most plants from mid-summer on. A winter-killed bud good enough for this purpose can often be found in spring. Thus one can identify a Willow as such at almost any time of year, and this is usually sufficient for this difficult genus.

MP
1

Typical "pussies," which become Willow flowers.

female (pistillate) flowers

male (staminate) flowers

fruit

fruit showing cotton

flower bud

leaf bud

Note: All Willow buds, regardless of shape or size, are covered by a single scale.

Note:
The twigs of Pussy Willow vary more among themselves than they do from the other three Willows shown here.

Dwarf Gray W.   Gray Willow   Silky Willow

These are all Pussy Willow twigs.

typical Willow bark

MP 2

168

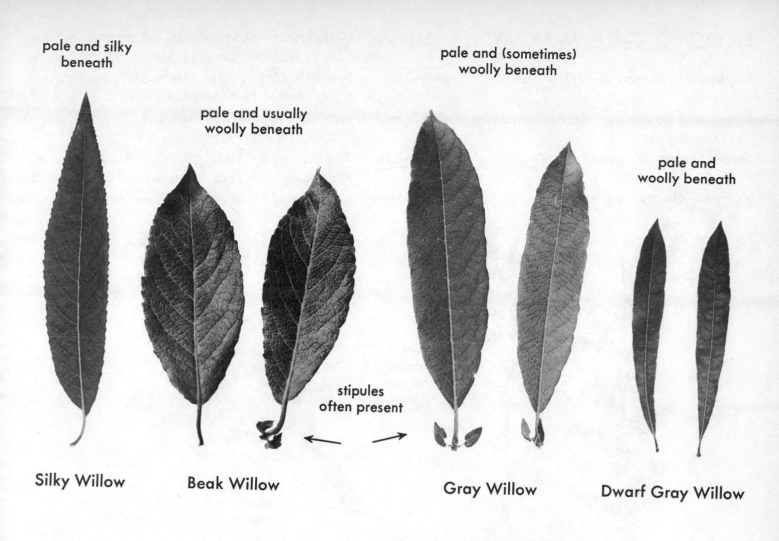

pale and silky
beneath

pale and usually
woolly beneath

pale and (sometimes)
woolly beneath

pale and
woolly beneath

stipules
often present

Silky Willow   **Beak Willow**   **Gray Willow**   **Dwarf Gray Willow**

green beneath;
not silky or woolly

pale beneath;
(usually) not silky or woolly

See also
Ground Covers,
MP 166A.

Shining Willow

These five leaves are all Pussy Willow
(all five forms are often found on the same plant).

Bearberry Willow

MP
3

BAYBERRY; GALE; WAX-MYRTLE—*Myrica*

Dioecious or Monoecious   Alternate shrubs with aromatic leaves, twigs and bark

Bayberry	*Myrica pensylvanica*	May-July	Nfd. to N. C., largely on coastal plain; locally w. to n. Ohio and w. N. Y. near Lake Erie 10 ft.
Evergreen Bayberry (Wax-myrtle)	*M. heterophylla*	April-June	coastal plain, s. N. J. to Fla. & La. 10 (occ. 15) ft.
Sweet Gale	*M. gale*	April-June	N. Amer.: s. to s. N. E., s. N. Y., n. Pa., Mich., Wis., Minn. & west; also occ. s. in mts. to N. C. & Tenn. 5-6 ft.
Wax-myrtle	*M. cerifera*	April-June	s. N. J. mostly near coast to Fla. & Tex. 35 ft.

male (staminate) flowers

female (pistillate) flowers

Bayberry
(Evergreen B. and Wax-myrtle similar)

waxy, berry-like fruit

Bayberry
(Evergreen B. and Wax-myrtle similar but slightly smaller)

barks typical of Bayberry, Evergreen B. and Wax-myrtle

Bayberry: warm brown, tan or light gray
Wax-myrtle: similar
Evergreen Bayberry: similar shape, but densely fuzzy and usually blackish green in winter

Aromatic leaves, all with typical Bayberry odor

evergreen (semi-evergreen
    in n. part of range)
small golden-yellow resin dots
    on *both* sides (best seen with lens)
narrow leaves

evergreen
resin dots beneath only
pale and dull beneath

deciduous (sometimes lasting
    into winter)
resin dots beneath only
yellow green and shiny beneath

Wax-myrtle

Evergreen Bayberry

Bayberry

winter
catkins

golden-brown
male (staminate)
flowers

reddish
female
(pistillate)
flowers

fruit

MP
5

Sweet Gale

Sweet Gale is considered a subgenus. The leaves are similar to the Bayberries and Wax-myrtle, and all are aromatic shrubs, but notice that Sweet Gale is the only one with winter catkins and that its flowers, fruit, and twigs are quite different from the others.

ALDER—*Alnus*   Monoecious   (Flowering dates are important)   Alternate

Mountain Alder *Alnus crispa* (American Green Alder)	June-July	Lab. to Alaska, s. to Mass. & w. to Minn.; also appears in mts. of w. N. C.	10 ft.
Sea-side Alder *A. maritima*	Aug.-Sept.	Del. & Md., also Okla.	30-35 ft.
Smooth Alder *A. rugosa*	Feb.-May	N. S. & Me., N. Y., Ohio, Ind., Ill., Mo. & Okla., s. to Fla. & La.	15 (occ. 25) ft.
Speckled Alder *A. incana*	March-May	N. S. to Minn., s. to s. N. E., w. to W. Va., n. Ill. & n.e. Ia.	15 (occ. 25) ft.

Note: Both the Smooth and Speckled Alders grow side by side in much of their range. Hybridization occurs frequently, making identification confusing. However, when true to type, the differences are evident; when not, it is best to suspect a hybrid, or intermediate form. The other two Alders shown here are easy to identify at any time of year.

female (pistillate) flowers

developing into fruit (seed-bearing cones)

seed without wings

conspicuous thin wings on seed

seeds x 3

Mountain Alder

other Alders

male (staminate) flowers

typical Alder flowers

Mountain Alder   Sea-side Alder   Smooth Alder (Speckled A. similar)

Mountain Alder   Sea-side Alder   Smooth Alder   Speckled Alder female (pistillate) catkins as seen in winter abruptly drooping

Speckled Alder

← male (staminate) catkins

(Immature) male (staminate) catkins; no female (pistillate) catkins appear in winter on the Mountain Alder, but are produced in spring from a winter bud. →

female (pistillate) catkins *not* drooping

Smooth Alder

twigs typical of Smooth and Speckled Alders

← male (staminate) catkins; (similar to Speckled Alder at top)

Mountain Alder

male (staminate) catkins →

female (pistillate) catkins →

No catkins will be found in winter on the Sea-side Alder, as it flowers in late summer from catkins produced the same summer. The picture at right shows these catkins as seen in August before flowering; also, immature fruit produced from flowers the previous summer.

Sea-side Alder

Sea-side Alder

MP 7

Mountain Alder

Speckled Alder

Sea-side Alder

Smooth Alder

MP
8

BIRCH—*Betula*   Monoecious   Alternate

Dwarf Birch	*Betula glandulosa*	June-Aug.	Nfd. to Alaska, s. to mts. of Me., N. H. & N. Y.; also appears in Mich. & some w. states	6 ft.
Low Birch	*B. pumila*	May-June	Nfd. & west, s. to Conn., Ill. & Ia.	15 ft.

male (staminate) flowers

winter catkins typical of Birches

female (pistillate) flower

typical Birch flowers

typical Birch fruit

conspicuously glandular-dotted

usually hairy (use lens)

(not glandular-dotted)

Low Birch

Dwarf Birch

bark typical of both these Birches

Low Birch

Dwarf Birch

MP
9

175

HAZELNUT or HAZEL—*Corylus* Monoecious Alternate

American Hazelnut	*Corylus americana*	April-May	Me. to Sask., s. to Ga. & Okla.	10 ft.
Beaked Hazelnut	*C. cornuta*	April-May	Nfd. to B. C., s. to Ga., Mo. & Colo.	10 ft.

Winter catkins develop into male flowers shown below.

twigs largely hairless; tips of buds whitish

long male (staminate) flowers

short male (staminate) flowers

twigs distinctly hairy (hairs sometimes wear off by end of winter)

red female (pistillate) flowers

Hazelnuts are in the same genus as Filberts, and are equally good to eat.

American Hazelnut

Beaked Hazelnut

MP 10

leaf-stalk
fuzzy →

American Hazelnut

typical Hazelnut barks

← leaf-stalk
not fuzzy

Beaked Hazelnut

SWEET FERN—*Comptonia peregrina* Monoecious April-May Alternate
Lab. to Alaska, s. to N. C., Tenn., Ill., Wis., Minn. & west 5 ft.

male
(staminate)
flowers

female
(pistillate)
flowers

winter catkins

burr-like
fruit

aromatic leaves and twigs

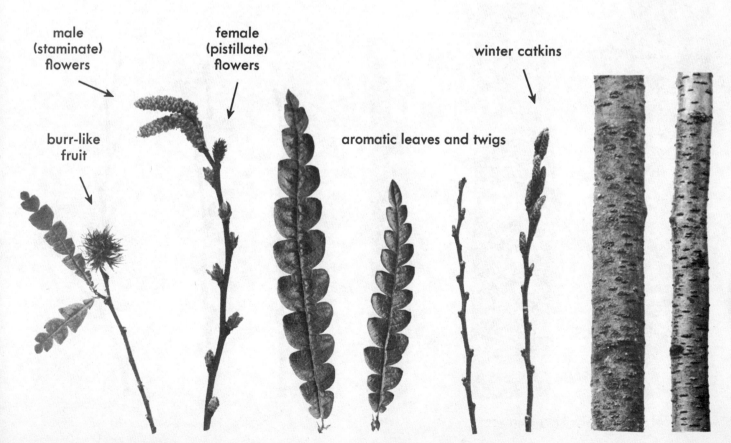

AMERICAN HORNBEAM — *Carpinus caroliniana* Monoecious April-May Alternate
(Blue Beech; Ironwood)
N.S. to Minn., s. to Fla. & Tex. 35 ft.

reddish
female
(pistillate)
flowers

male (staminate)
flowers

winged seeds

typical muscular-looking stems

MP
12

CHESTNUT — *Castanea*   Monoecious   Alternate

Chinquapin   *Castanea pumila*   June   s. N. Y., e. Pa., W. Va., Tenn., s. to Fla. & Tex.

15-30 ft.

female
(pistillate)
flowers

male
(staminate)
flowers

Only a single nut is
produced in each burr.

OAK—*Quercus* Monoecious Alternate

Chinquapin Oak *Quercus prinoides* April-May Me. to Minn., s. to Ala. & Tex.
  White Oak Group 10 (occ. 15) ft.
Scrub Oak *Q. ilicifolia* April-May Me. to N. Y., s. to N. C. & W. Va.
  Red (or Black) Oak Group 18-20 ft.

Note: There are two main subgenera: (1) the White Oak Group, with acorns that mature in one year and leaves with lobes that may be rounded or sharp-pointed but never bristle-tipped; (2) the Red (or Black) Oak Group, with acorns that require two years to mature and leaves with bristle-tipped lobes. Considerable hybridization occurs among different species within the same group, and this must be borne in mind when identifying Oaks, as they will not always appear true to type.

bristle-tipped lobes

Note clusters of buds
at ends of twigs,
typical of all Oaks.

Scrub Oak

typical Oak flowers

Chinquapin Oak

Scrub Oak

MP
14

bristle-tipped lobes

lobes (teeth) usually sharp-pointed
but *not* bristle-tipped

Scrub Oak

Chinquapin Oak

Scrub Oak

Chinquapin Oak

MP
15

BUFFALO-NUT — *Pyrularia pubera*
(Oil-nut)

Sub-dioecious   May   Alternate
Penn. to Ga. & Ala.   12 ft.

The Buffalo-nut is parasitic on the roots of other shrubs and trees, but appears to be a separate shrub.

Note: The twigs fall off easily, leaving round twig-scars above the leaf-scars.

twig-scar

leaf-scar

MP
16

AMERICAN MISTLETOE— *Phoradendron flavescens* Dioecious Oct.-Dec. Opposite
N. J. to Kans., s to Fla. & Tex.; evergreen;
parasitic on branches of various trees   3 ft.

inconspicuous
flowers

fruit white, sticky
when crushed;
carried from tree
to tree by birds

stems green; twigs brittle
at base, falling off and
leaving round twig-scars

JAPANESE KNOTWEED—*Polygonum cuspidatum* Perfect or Polygamous   Aug.
Asia; commonly escaped elsewhere   6-8 ft.   Alternate

This is not a true woody shrub, but a perennial plant.
It is included here because it is often mistaken for Bam-
boo with its hollow, woody-appearing stems. These die
back to the ground during the winter, new ones coming
up each spring.

MP
17

183

BARBERRY—*Berberis* Perfect Alternate

American Barberry	B. canadensis	May-June	s. Pa. & W. Va., s. to Ga. & s.e. Mo.	3 (occ. 5) ft.
Common Barberry	B. vulgaris	May-June	Europe; planted here, commonly escaped	7-8 ft.
Japanese Barberry	B. thunbergii	April-May	Japan; planted here, commonly escaped	7-8 ft.

All Barberry flowers are yellow.

red

red

Common Barberry
(American B. similar,
but with fewer flowers
in each raceme)

Common
Barberry

red

American Barberry

Japanese Barberry

this leaf a
juvenile
form

Common
Barberry

American Barberry

The inner bark of all Barberries is yellow.

Twigs and young stems are reddish brown in winter; older stems are gray (top row).

thorns curving, mostly triple
(except sometimes single on small twigs)

thorns straight and slender, mostly single
(except occasionally triple on largest stems)

American Barberry

Japanese Barberry

Twigs and all older stems are uniformly gray in winter (bottom row).
Thorns are mostly triple (or more),
except sometimes single on small twigs.

Common Barberry

MP
19

SWEET SHRUB—*Calycanthus floridus*   Perfect   April-July   Opposite
(Carolina Allspice; Hairy Strawberry Shrub)   Pa. & W. Va. to Fla. & Miss.   9 ft.

Note: The Smooth Strawberry Shrub—*C. fertilis*—is found in much the same range, the principal differences being that the flowers are less fragrant and the leaves relatively hairless beneath.

leaves downy beneath; aromatic

flowers red brown; strawberry-scented

twigs and bark aromatic

SPICE-BUSH—*Lindera benzoin*
(Spiceweed; Benjamin Bush)

Dioecious or Polygamous March-April Alternate
Me. to Ont. & Mich., s. to Fla. & Tex.   15 ft.

flowers
light yellow

berries yellow
to red

leaves aromatic
when crushed

bark and twigs aromatic

leaf buds
(growth buds)

flower buds

in clusters,
alternately
along stems

MP
21

MOCK-ORANGE—*Philadelphus*   Perfect   Opposite

Large-flowered Mock-orange	*P. grandiflorus*	May-June	Va. & Tenn., s. to Fla. & Miss.; commonly planted in n. states 9 ft.
Scentless Mock-orange	*P. inodorus*	May-June	Va. & Tenn., s. to Fla. & Miss.; commonly planted in n. states 9 ft.
Sweet Syringa (Mock-orange)	*P. coronarius*	May-June	s. Europe; commonly planted here 9 ft.

note: usually three blossoms

note: usually more than three blossoms

very fragrant

Large-flowered Mock-orange
(Scentless M. similar)

Sweet Syringa

note: triangular leaf-scar

Scentless Mock-orange
(Large-flowered M. similar)

twigs and bark typical of Mock-oranges

Sweet Syringa

MP
22

mostly without teeth

mostly with teeth

Scentless Mock-orange

Sweet Syringa
(Large-flowered M. similar)

VIRGINIA-WILLOW—*Itea virginica* Perfect May-June Alternate
(Tassle-white) N. J., e. Pa. & Va. to Okla.,
s. to Fla. & Tex. 4 (occ. 9) ft.

young stems and twigs
green, tinged with red

Leaf-stalk often re-
mains on twig when
main part of leaf
falls off.

FALSE SPIREA—*Sorbaria sorbifolia* Perfect June-July Alternate
(Mountain Ash Spirea)
n. Asia; planted here & escaped   6 ft.

SHRUB YELLOW-ROOT—*Xanthorhiza simplicissima*

Polygamous   April-May   Alternate
N. Y. to W. Va., s. to Fla. & Ala.   2 ft.

flowers
brownish purple

roots and inner bark
of stems yellow

MP
25

DEUTZIA—*Deutzia scabra* Perfect June Opposite
Japan; escaped elsewhere 7-8 ft.

stems hollow
except at
nodes (joints)

MP
26

NINEBARK—*Physocarpus opulifolius* Perfect   June   Alternate
Que. to Minn. & Colo., s. to S. C. & Ark.   9 ft.

Fruit becomes red,
then greenish,
finally brown.

Stems have
solid pith.

HYDRANGEA — *Hydrangea*   Flowers of two kinds: Perfect and Sterile   Opposite

Peegee Hydrangea   *H. paniculata grandiflora*   Mostly sterile flowers   Aug.-Sept.
Japan & China 15-20 (occ. 30) ft.

Wild Hydrangea   *H. arborescens*   Mostly perfect flowers with a few marginal sterile ones   June-July
(Seven-bark)   N. Y. to s. la., s. to Fla. & La. 3-4 (occ. 10) ft.

Note: A variety of Wild Hydrangea known as Hills of Snow — *H. a. grandiflora* — has mostly sterile flowers.

red leaf-stalks →

Peegee Hydrangea

Wild Hydrangea

Peegee Hydrangea

typical
Hydrangea twigs

Wild Hydrangea

MP
28

pyramidal flower clusters;
flowers mostly sterile

Peegee Hydrangea

As most of the flowers are
sterile, there is little actual
fruit, but the flower remnants
last a considerable time.

flat flower cluster; flowers
mostly perfect, with only a
few sterile marginal ones

MP
29

As most of the flowers are
perfect, fruit capsules usually
develop and can be found
much of the year.

Wild Hydrangea

CURRANT and GOOSEBERRY—*Ribes* Perfect Alternate
(Dioecious in some foreign species not included here)

I. GOOSEBERRIES: flowers (and fruit) few (one to five) from any one point on stems

European Gooseberry *R. grossularia*	May-June	Europe, n. Africa & Asia; escaped here	3 ½ ft.
Prickly Gooseberry *R. cynosbati*	May-June	N. B. to Man., s. to N. C., n. Ala. & Mo.	6 ft.
Smooth Gooseberry *R. hirtellum* (Low Wild Gooseberry)	April-July	Lab. to Man., s. to N. C. & Tenn.	3 ft.

Note: Considerable confusion exists concerning the actual differences between the Smooth Gooseberry and the Round-leaved Gooseberry — *R. rotundifolium* — and until further study indicates two clear-cut species, it would seem pardonable to consider them one species, with perhaps some local but unspectacular variations. The two are treated as one here, and the range given is for both types.

II. CURRANTS: flowers (and fruit) in racemes of five or more individual flowers (or fruit)

American Black Currant *R. americanum* (Wild Black Currant)	April-June	N. S. to Alb., s. to Va., Mo. & west	6 ft.
Bristly (Black) Currant *R. lacustre*	May-Aug.	Nfd. to Alaska, s. to n. N. E., w. Mass., N. Y., Pa., n. Ohio, Mich., Wis., Minn. & west	3 ft.
European Black Currant *R. nigrum*	April-June	Europe & Asia; escaped here	6 ft.
Garden Red Currant *R. sativum*	April-June	Europe; escaped here	5 ft.
Skunk Currant *R. glandulosum*	May-Aug.	Lab. to B. C., s. to s. N. E., N. Y., Pa., W. Va., mts. of N. C., Ohio, Mich., Wis. & Minn.; stems reclining or creeping, with ascending branches	2 ft.
Note: Most parts of this plant have skunk-like odor when crushed, especially the twigs.			
Swamp Red Currant *R. triste*	April-June	Lab. to Alaska, s. to s. N. E., N. Y., Pa., W. Va., Ill., Ia. & west; arching or straggling stems	2 ½ ft.

I. GOOSEBERRIES: flowers few (one to five) from any one point on stem

ovary bristly →

stamens short (not protruding)

ovary smooth (not hairy or bristly) ↙

ovary covered with soft hairs ↓

stamens long and easily seen

flowers greenish yellow

flowers greenish, tinged with pink or red

Prickly Gooseberry   **Smooth Gooseberry**   **European Gooseberry**

MP 30

196

II. CURRANTS: flowers in racemes (of five or more individual flowers)

A. flowers flat, saucer-shaped

racemes drooping

racemes erect

ovary bristly

ovary smooth

Bristly Currant
(flowers pink to purplish red)

Garden Red Currant
(flowers greenish yellow)
(Swamp Red C. similar,
but flowers salmon or dull
pink)

Skunk Currant
(flowers white or with
slight pinkish tinge)

B. flowers bell-shaped

American Black Currant
(flowers pale yellow)

European Black Currant
(flowers greenish, tinged with red or pink)

MP
31

I. GOOSEBERRIES

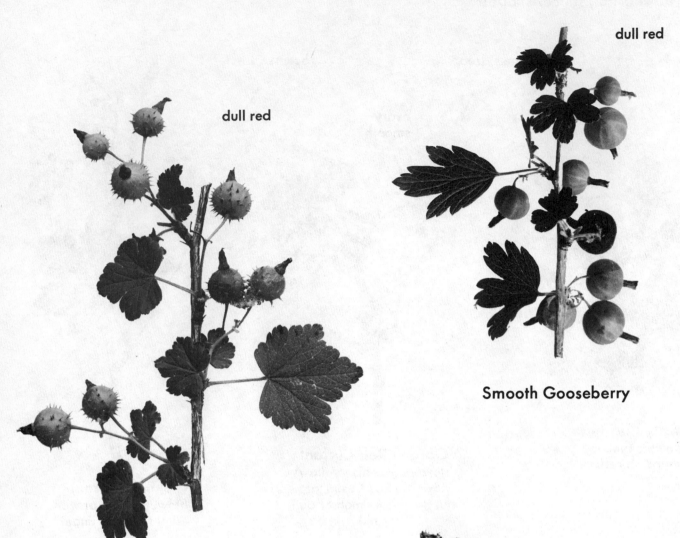

dull red

dull red

Smooth Gooseberry

Prickly Gooseberry

red, yellow or green

European Gooseberry

MP
32

II. CURRANTS

bristly berries

black

Bristly Currant

bright red

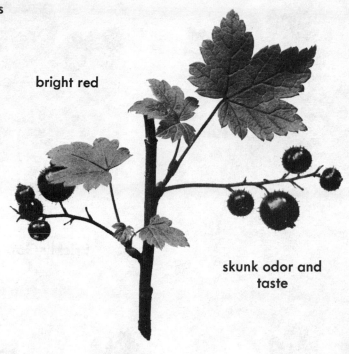

skunk odor and taste

Skunk Currant

smooth berries

shiny black

American Black Currant
(European Black C. similar)

bright red

Garden Red Currant
(Swamp Red C. similar,
except slightly smaller,
harder and less juicy)

MP
33

leaves
velvety
underneath

Prickly Gooseberry

Smooth Gooseberry

teeth
generally
more rounded
than those of
Smooth Gooseberry

European Gooseberry

MP
34

II. CURRANTS

Bristly Currant

hairy
leaf-stalk

Skunk Currant

Swamp Red Currant

reddish
leaf-stalk

resin-
dotted both sides
(more clearly so
beneath; use lens)

American Black Currant

Garden Red Currant

resin-
dotted
beneath only

European Black Currant

All Gooseberries are armed (have some thorns or bristles).

Prickly Gooseberry

Smooth Gooseberry

The Bristly Currant is the only armed Currant. The small stems and twigs are golden brown and densely bristly.

MP
36

European Gooseberry

Bristly Currant

All Currants are unarmed (no thorns or bristles) with the exception of the Bristly Currant shown on the opposite page.

American Black Currant

Garden Red Currant

Stems and twigs have
skunk odor when broken.

brown buds

red buds

European Black Currant

Skunk Currant

Swamp Red Currant

MP
37

WITCH-HAZEL—*Hamamelis virginiana* Perfect Sept.-Nov. Alternate
Que. to Minn., s. to Ga. & Ark. 15 ft.

pale yellow flowers

Fruit capsule bursts open when ripe, expelling seeds for wide distances (up to 50 ft.).

naked buds (without scales)

MP
38

CHOKEBERRY—*Aronia* Perfect April-May Alternate

Black Chokeberry   *Aronia melanocarpa*   N. S. to Ont., s. to S. C., Tenn. & la.   3-5 ft.
Red Chokeberry   *A. arbutifolia*   s. N. E., Ont. & Ohio, s. to Fla. & Tex.   10 ft.

Note: Purple Chokeberry — *A. prunifolia* — N. S. to Ont. & Mich., s. to Va. & Ill., appears to be an intermediate species, with characteristics between the other two Chokeberries. The fruit is dark red or purple; the height is variable, but up to 10 ft.

typical Chokeberry
flowers

Ripe berries of Black C. are black, those of Red C. are red.

Note: There is a row of black hair-like glands on the midrib on the upper side of all Chokeberry leaves (best seen with a lens).

typical Chokeberry leaves
Black C.: smooth beneath
Red C.: fuzzy beneath

Red Chokeberry
(fuzzy twigs)

Black Chokeberry
(smooth twigs, often with a silvery coating)

typical Chokeberry
barks

MP
39

205

SPIREA — *Spiraea* Perfect Alternate

Corymbed Spirea	*Spiraea corymbosa*	June-July	s. Pa. to Ga. & Ky.	3 ft.
(Birch-leaved Meadowsweet)				
Hardhack	*S. tomentosa*	July-Sept.	N. S. to Ont. & Minn., s. to Ga. & Ark.	
(Steeplebush)				3-4 ft.
Meadowsweet	*S. latifolia*	June-Aug.	Nfd. to Mich., s. to N. C.	4-5 ft.
(Broad-leaved Meadowsweet)				

Note: There are a large number of foreign Spireas much used by nurseries in this country, but few, if any, have become naturalized.

pink

white

flat flower cluster

pale beneath; often rusty and woolly

green beneath

usually no teeth below the middle

Hardhack   Meadowsweet   Corymbed Spirea

MP 40

Hardhack

Meadowsweet

Corymbed Spirea

flat fruit cluster

The twigs of Hardhack are woolly; the others have smooth twigs.

typical Spirea barks

Older barks are gray; young shoots are brown, sometimes smooth and shiny, at other times somewhat shaggy.

typical Spirea twigs

MP 41

FIRETHORN—*Pyracantha coccinea lalandii* Perfect   May-June   Alternate
Evergreen   s. Europe & w. Asia; naturalized Pa. & south, hardy farther north   8 (occ. 20) ft.

orange-red berries

usually very
thorny

HAWTHORN—*Crataegus*   Perfect   mostly April-June   shrubs or trees   Alternate

There are hundreds of species in the genus, at one time over one thousand being named in North America alone. However, many of these are no longer considered to be separate species, but rather hybrids, varieties or forms.

Note variation in leaf shapes.

usually very thorny

round little red buds, usually at the ends of small spurs

SHADBUSH—*Amelanchier*
(Juneberry; Service-berry)

Perfect   April-May   shrubs or small trees   Alternate
N. Amer.: s. to Mexico

This is a large and confusing genus; there is considerable difference of opinion concerning the various species.

Fruit turns from pink
or red to purple or black.

MP
44

NEW JERSEY TEA—*Ceanothus americanus* Perfect   June-Sept.   Alternate
Canada, s. to Fla. & Tex. 3 ft.

fruit at first covered
by black case, opening
to expose three white nutlets

JETBEAD—*Rhodotypos scandens* Perfect   May-June   Opposite
Japan, China; escaped elsewhere 6 ft.

black berries

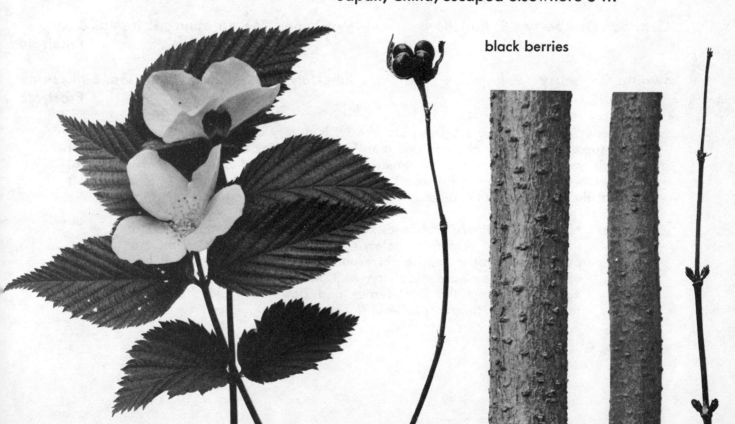

MP
45

BRAMBLE (Raspberries, Blackberries, Dewberries) — *Rubus*   Perfect   Alternate

I. RASPBERRIES

American Red Raspberry	*Rubus idaeus strigosus*	May-June	Nfd. to B.C., s. to Va., Tenn. & Wyo.; stems erect (not arching or rooting at tips) 6 ft.

(This is considered a variety of the European Red Raspberry—*R. idaeus*.)

Black Raspberry *R. occidentalis* (Black-cap; Thimbleberry)		May-June	N.B. to Minn., s. to Ga., Tenn. & west; stems (canes) arching up to 10 ft. long and rooting at tips
Purple Flowering Raspberry *R. odoratus* (Thimbleberry)		June-Aug.	N.S. to Mich., s. to Tenn. & Ga.; stems erect 6-8 ft.
Wineberry *R. phoenicolasius*		June-July	e. Asia; escaped in e. U.S.; arching stems to 10 ft. long and rooting at tips

II. BLACKBERRIES and DEWBERRIES

Highbush Blackberry *R. allegheniensis* (Allegheny Blackberry)		May-July	N.S. to Minn., s. to N.C. & Ark. 10 ft.
Smooth Blackberry *R. canadensis* (Thornless B.; Canada B.)		June-July	Nfd. to Ont. & Minn., s. to Ga., Tenn. & Ill. 10 ft.
Common Dewberry *R. flagellaris*		May-June	Me. to Minn., s. to Va. & Mo. Prostrate
Swamp Dewberry *R. hispidus*		June-July	N.S. to Mich., s. to Ga. & Ill.; semi-evergreen Prostrate

This genus, especially the subgenus Blackberry, is a very confusing one, as it is undergoing rapid evolutionary changes and therefore the species are not stable. It is thought that this variability is a result of extensive clearing of land and other environmental changes beginning with the arrival of the white man. However, as with other difficult genera, the main types are sufficiently clear-cut for practical purposes.

Note: The fruit of Raspberries separates easily from what is called the receptacle. A definite hole at the base of the fruit marks the point of separation, and a corresponding projection or knob is left on the stem (see Black Raspberry fruit, MP 50). This distinguishes all Raspberries from the Blackberries and Dewberries, which do not separate from the receptacle in this manner.

MP
46

I. RASPBERRIES
(see also next page)

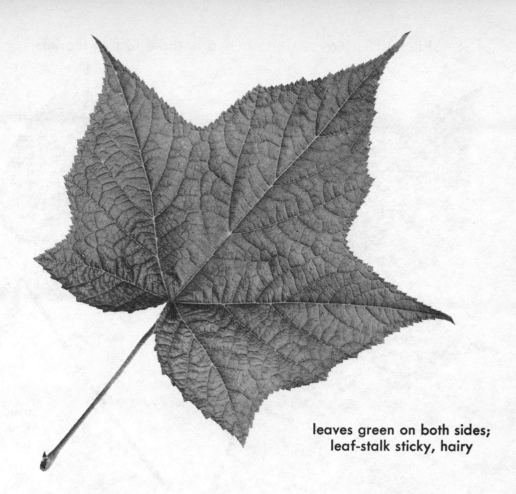

leaves green on both sides;
leaf-stalk sticky, hairy

Purple Flowering Raspberry

leaves white beneath;
leaf-stalk sticky, hairy

Wineberry

I. RASPBERRIES: leaves white beneath; three to five leaflets

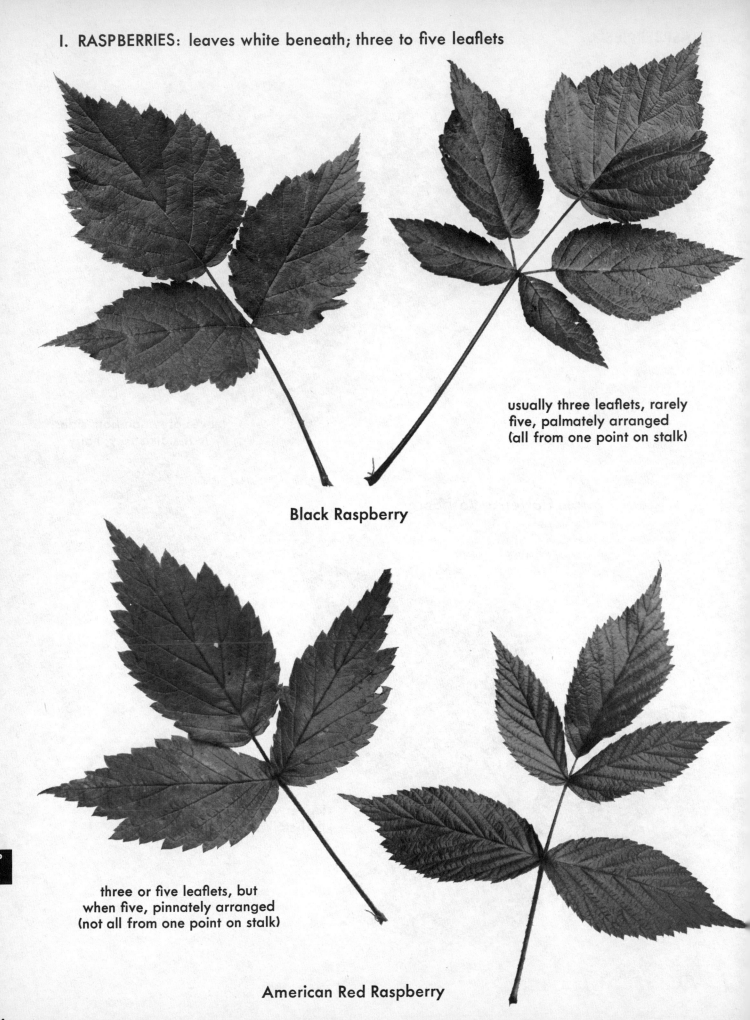

usually three leaflets, rarely five, palmately arranged (all from one point on stalk)

Black Raspberry

three or five leaflets, but when five, pinnately arranged (not all from one point on stalk)

MP 48

American Red Raspberry

II. BLACKBERRIES and DEWBERRIES:
leaves green beneath; three or five leaflets; when five, always palmately arranged

densely covered with woolly hairs on underside, velvety to touch

Highbush Blackberry
(Smooth B. similar but not velvety beneath)

generally very glossy; semi-evergreen, but often with reddish cast, especially in fall and winter; usually three, rarely five leaflets

three to five leaflets

MP 49

Swamp Dewberry

Common Dewberry

I. RASPBERRIES

magenta, turning
lavender pink

Black Raspberry
(Red R. similar)

**Purple Flowering
Raspberry**

Note white projec-
tion left after fruit
is removed (typical
of Raspberries).

Black Raspberries
are red before ripening,
but black when ripe.

Red Raspberries
are red when ripe.

red; rather
tasteless

Black Raspberry
(Red R. similar)

**Purple Flowering
Raspberry**

stems and
parts of flowers
(and fruit) densely
red-hairy; hairs
gland-tipped and sticky

red; rather
tasteless

Wineberry

II. BLACKBERRIES and DEWBERRIES

Swamp Dewberry
(Common D. similar, but petals usually larger)

Highbush Blackberry
(Smooth B. similar, but flower stems hairless)

fruit sour; black when ripe

hairy stems

Swamp Dewberry

fruit sweet; black when ripe

fruit sweet; black when ripe

Common Dewberry

Highbush Blackberry
(Smooth B. similar)

MP 51

I. RASPBERRIES

arching; rooting at tips; heavy white bloom over red stems

erect or arching; not rooting at tips; slight (occ. heavy) bloom; brown or red stems

Black Raspberry

American Red Raspberry

erect; not rooting at tips

arching; rooting at tips; densely red-hairy, with some thorns

Purple Flowering Raspberry

Wineberry

MP
52

II. BLACKBERRIES and DEWBERRIES: stems green or red without bloom

Blackberries: large; upright or arching

Dewberries: prostrate, or scrambling over low objects

stems round, or often grooved

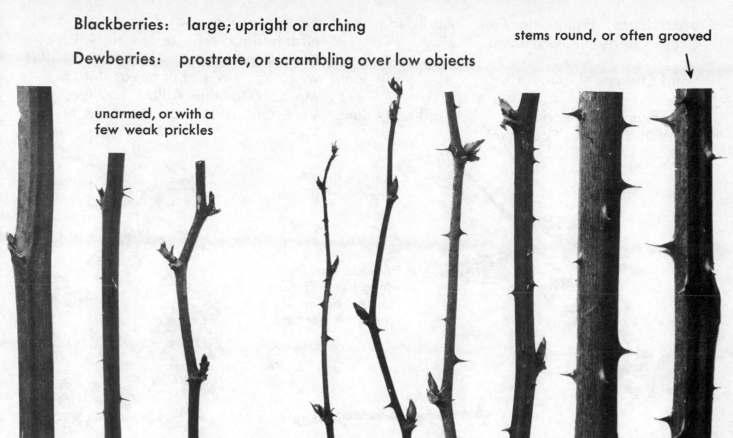

unarmed, or with a
few weak prickles

Smooth Blackberry

Highbush Blackberry (stout thorns usually present)

semi-evergreen;
usually very bristly slender stems

Swamp Dewberry

deciduous;
not bristly, but with a few stiff prickles

Common Dewberry

MP
53

CHERRY and PLUM—*Prunus*   Perfect   Alternate

Beach Plum	*Prunus maritima*	April-June	Me. to Va. near coast	6-8 ft.
Choke Cherry	*P. virginiana*	May	Nfd. to Sask. & S. Dak., s. to N. C., Kans. & west	30 ft.
Sand Cherry	*P. pumila*	April-early June	w. N. Y. to Wis., but in var. Me. & Mass., w. to Minn. & Ill.	3 (occ. 8) ft.
Wild Plum	*P. americana* (Wild Yellow or Red Plum)	April-early June	Vt. & Ont. to Minn., s. to Fla. & N. Mex.	25 ft.

flowers (and fruit) in terminal racemes

flowers (and fruit) along stems; (few or many, but not in terminal racemes)

Choke Cherry

red, turning blue black

Choke Cherry

Sand Cherry

purple or black

Wild Plum

(Beach P. and Sand Cherry similar)

red or purple

Beach Plum

red, rarely yellow

Wild Plum

MP 54

Note: Black Cherry, a tree, is sometimes confused with Choke Cherry, but the leaves of Black Cherry have rounded teeth, are glossier and usually narrower than those of Choke Cherry.

teeth usually few or missing along bottom third of leaf

Choke Cherry

Sand Cherry

Wild Plum

Beach Plum

MP 55

Twigs and bark have an unpleasant smell when broken.

tent-caterpillar egg mass, often found on Cherries and Plums

usually somewhat thorny (see also Thorn Key, p. 2)

Choke Cherry   **Sand Cherry**   **Beach Plum**   **Wild Plum**

bark usually not scaly until larger than this

Choke Cherry   **Beach Plum (Sand Cherry similar)**   **Wild Plum**

ROSE—*Rosa*   Perfect   Alternate

Multiflora Rose	*R. multiflora*	June-July	e. Asia; widely planted, escaped; arching shrub to 8 ft., rambling if support is available to 10-12 ft.
Pasture Rose (Carolina Rose)	*R. carolina*	June-July	N.S. to Minn., s. to Fla. & Tex.; small shrub 3 ft.
Prairie Rose	*R. setigera*	June-July	c. N.Y. to Mich., Ill. & Kans., s. to Fla. & Tex.; arching shrub to 6 ft., clambering if support is available, but not to the extent of the Multiflora R.
Rugosa Rose	*R. rugosa*	May-Aug.	e. Asia; widely planted, escaped 6 ft.
Smooth Rose (Meadow Rose)	*R. blanda*	May-July	Nfd. to Man., s. to n. & w. N.E., N.Y. & Pa., w. to Mo. 6 ft.
Swamp Rose	*R. palustris*	June-Aug.	N.S. to Minn., s. to Fla. & Ark. 8 ft.
Sweet-Brier (Eglantine)	*R. eglanteria*	May-July	Europe; naturalized here 10 ft.
Virginia Rose	*R. virginiana*	June-July	Nfd. to s. Ont., N.Y. & N.E., s. to N.C., Ala., Tenn. & Mo., but apparently rare or absent in states w. & n. of Pa. 6 ft.

The Roses comprise a beautiful but confusing group of plants. There have been many differences of opinion concerning the various species, no two experts ever totally agreeing on the relative importance of apparently conflicting factors. In addition to man-made confusion, the Roses themselves contribute their own peculiar ways to complicate the situation. As a group they have distinctive features that make them known at once as Roses, but the individuals, within many species, appear to be deliberately perverse, and sports, variations and hybrids are common, making for instability within species. It is important to bear this in mind when looking for acceptable features on which to base species identification.

There are certain characteristics, nevertheless, that can be depended upon to remain relatively stable and which can be used to divide the Roses into distinct groups. It is best to attempt species identification beginning at flowering time and continuing through the fruiting period, although on better acquaintance it is often possible to identify many Roses at any season. As with a few other genera included in this book, main groupings are the primary objectives, and representative species of only the most common types are shown.

Always look at all available details; a combination of details, not just one, is essential for the proper identification of Rose species.

MP
57

223

Note: The pistils (female part of flower) of Multiflora Rose and Prairie Rose are long, forming a protruding column from the center of the flower.

pink to white

white

Note: Both of these are Rambler Roses; the others are erect or arching but do not clamber.

Prairie Rose

Multiflora Rose

note pistils not extended; centers of flowers flat

magenta

pink to white

Smooth Rose

(Pasture R., Swamp R., Sweet-Brier and Virginia R. similar)

Rugosa Rose

MP 58

Note: Roses bear a red fruit called a "hip." This is distinctive of the genus and in some cases helpful in distinguishing the species. The Rugosa and Smooth Roses retain their sepals (leaf-like appendages at the end of the hip); the others lose their sepals by the time the fruit is ripe, or earlier.

sepals persistent
after fruit ripens

Rugosa Rose

Prairie Rose

Multiflora Rose

Smooth Rose

Sweet-Brier

Pasture Rose

Virginia Rose

Swamp Rose

MP
59

Note: The fruit of these four Roses vary considerably and should not be used to determine the species. They are shown here to indicate characteristics possible for any one.

Note: Rose leaves bear stipules (leafy growths attached to the leaf-stalk) characteristic of the genus and in some cases typical of a particular species (see Multiflora and Swamp Roses).

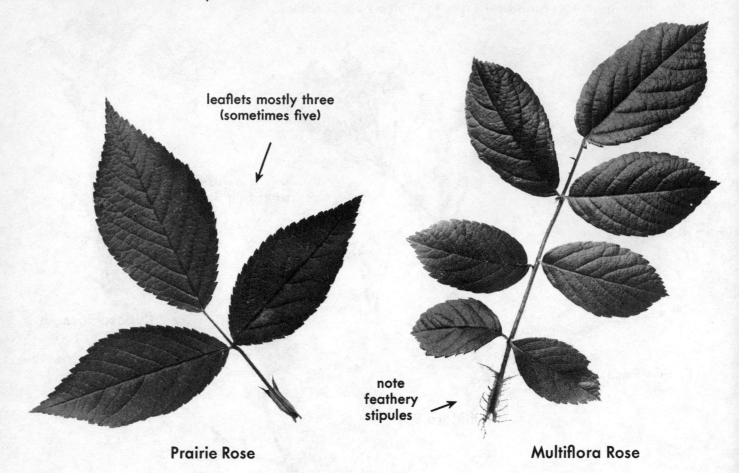

leaflets mostly three (sometimes five)

note feathery stipules

Prairie Rose

Multiflora Rose

Pasture Rose (Virginia Rose similar)

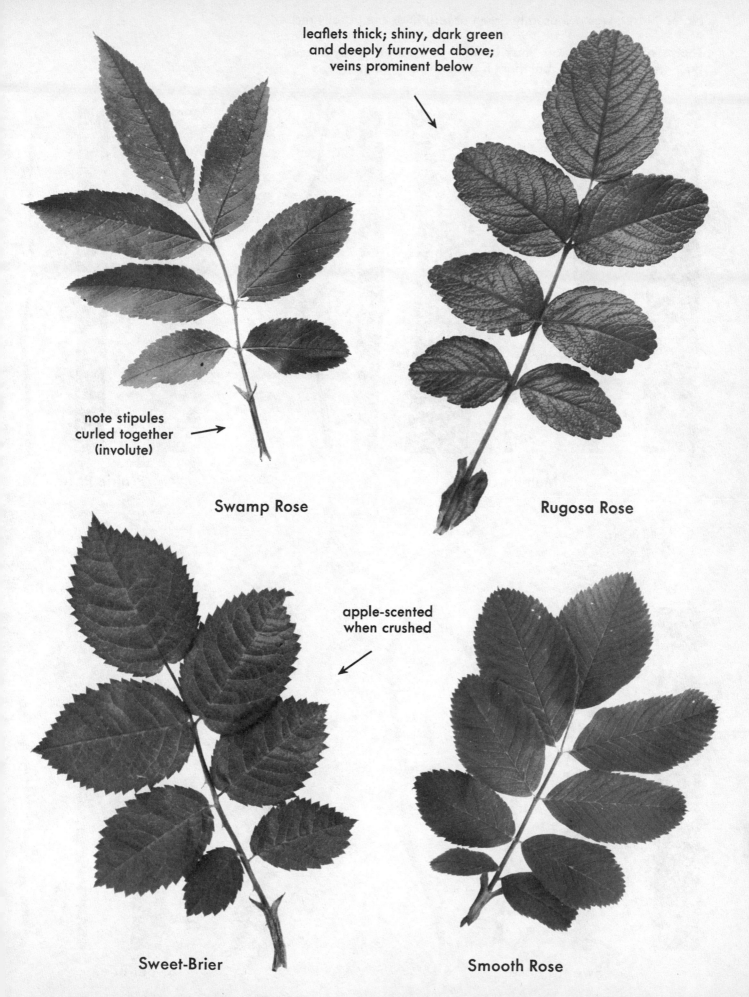

leaflets thick; shiny, dark green
and deeply furrowed above;
veins prominent below

note stipules
curled together
(involute)

Swamp Rose

Rugosa Rose

apple-scented
when crushed

Sweet-Brier

Smooth Rose

Note: Rose twigs are usually green or red; buds are usually red.

Thorns of Multiflora Rose may be light-colored or nearly black; they are usually heavy, but there is a thornless variety.

Multiflora Rose

Prairie Rose

Swamp Rose

Sweet-Brier

rough, blackish bark

few if any thorns; weak bristles
on lower part of large stems only

Rugosa Rose

Smooth Rose

thorns mostly curved

thorns mostly straight, needle-like

Virginia Rose

Pasture Rose

MP
63

CINQUEFOIL—*Potentilla*   Perfect   Alternate

Shrubby Cinquefoil	*P. fruticosa*	May-Aug.	N. Hemisphere; in N. Amer., Arctic, s. to N. J., Pa., Ill., Ia., S. Dak., N. Mex. & west 2-3 ft.
Three-toothed Cinquefoil	*P. tridentata*	May-Sept.	Arctic, s. to N.E. & N.Y., w. to N. Dak., s. in mts. to Ga.; subshrub Prostrate

yellow flowers

white flowers

Shrubby Cinquefoil

Three-toothed Cinquefoil

MP 64

230

DYER'S GREENWEED — *Genista tinctoria*
(Woad Waxen; Whin)

Perfect   June-Aug.   Alternate
Europe & Asia; naturalized e. states;
semi-evergreen   3 ft.

yellow
flowers

twigs and small stems green

SCOTCH BROOM — *Cytisus scoparius*

Perfect   May-June   Alternate
Europe; naturalized e. states   6 ft.

yellow
flowers

twigs and small stems green

Well-developed leaves
have three leaflets, but
the upper leaves are
often reduced to one,
and sometimes few
leaves develop at all, the
stems being bare even
in summer.

FALSE INDIGO—*Amorpha* Perfect Alternate

Indigo Bush *A. fruticosa* (False Indigo)	May-June	Mich. to Minn., s. to Fla. & La.; planted and escaped in n.e. states	12 ft.
Lead-plant *A. canescens* (Shoestring)	June-July	Mich. to Sask., s. to Ind., Tex. & N. Mex.	3 ft.

purple flowers with golden pollen, typical of both species

Lead-plant

Indigo Bush

Lead-plant

Indigo Bush

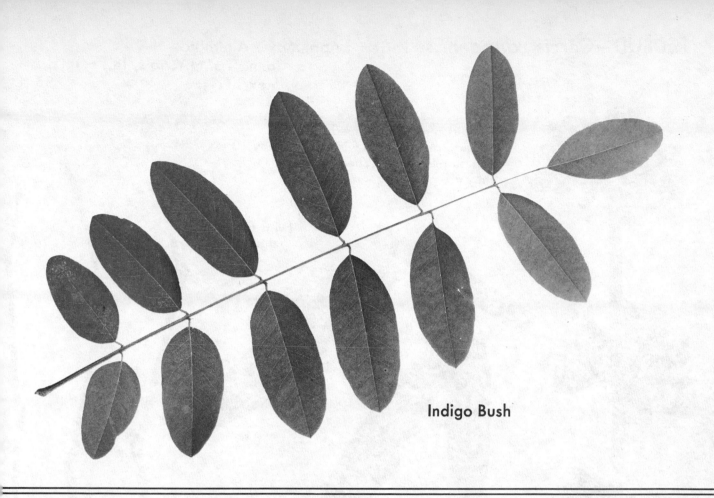

Indigo Bush

GORSE—*Ulex europaeus*   Perfect   March-June   Alternate
Europe; naturalized Atlantic states and
West Coast   3-(rarely) 6 ft.

yellow
flowers

black pods
covered with
white fuzz

REDBUD — *Cercis canadensis* Perfect April-May Alternate

s. Conn. & s. N.Y. to s. Ia., s. to Fla. & N. Mex.

35 ft.

Flowers remain unopened for a considerable time, at which stage they are purplish pink. When fully open they are pale pink.

twigs often zigzag

The flowers are borne on the older stems, not on the younger wood as with most shrubs. This feature is distinctive, and evidence of it is seen during most of the year, not only by the position of the flowers, but also those of the pods and clusters of flower buds. The twigs, as seen above, bear only inconspicuous growth buds.

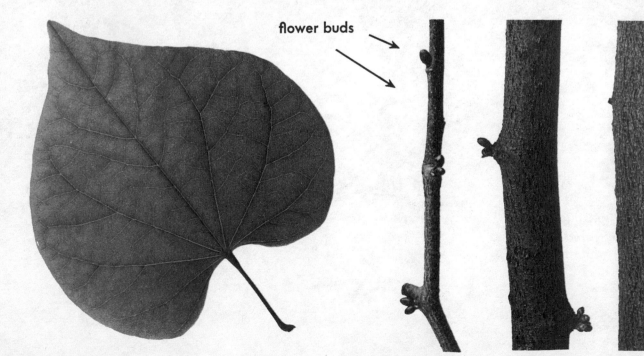

flower buds

BRISTLY LOCUST—*Robinia hispida* Perfect May-June Alternate
Va. to Ky., Ga. & Ala.; planted & escaped northward 3-4 ft. (occ. more)

pink flowers

PRICKLY ASH—*Zanthoxylum americanum*
(Toothache-tree)

Dioecious or Polygamous   April-May   Alternate
Que. to N. Dak., s. to Ga. & Okla. 25 ft.

greenish
flowers

fruit red, turning brown,
containing shiny black seeds

rust-colored buds

MP
70

HOP-TREE—*Ptelea trifoliata* Polygamous May-June Alternate
s.w. Que. & N.Y. to s. Ont. & Mich., s.
to Fla. & Tex.; escaped elsewhere 25 ft.

flowers
greenish white

SMOKE-TREE — *Cotinus*   Dioecious or Polygamous   Alternate

American Smoke-tree (Chittam-wood)	C. *americanus*	Dioecious	May-July	Tenn. & Mo. to Tex. & Ala. 35 ft.
Smoke-tree	C. *coggygria*	Polygamous	May-July	s. Europe, c. China & Himalayas; planted here & occ. escaped 15 ft.

tiny individual flower
stems covered with
pinkish hairs

flowers of
both species
yellowish

X →

American
Smoke-tree

Smoke-tree

small stems
covered with
tiny hairs

actual fruit

Note: These plants produce very few fertile
flowers. The individual flower stems of the
sterile flowers are covered with fine pink
or brownish hairs. These stems continue to
grow after flowering and become the
spectacular plumes that give these plants the
name of Smoke-trees. Notice that the
flower of American Smoke-tree (upper left)
shows the beginnings of the hairy stems, and
the picture of the fruiting plume (lower left)
shows these stems much elongated, with
only a few of them bearing actual fruit.

MP
72

A small section of a large fruiting plume, typical of
both species. This developed from a flowering section
such as shown above at X. Bear in mind that both the
flowering and fruiting sections are actual size.

American
Smoke-tree

Smoke-tree

barks typical of both
species when small

American Smoke-tree

Smoke-tree

SUMAC—*Rhus* Dioecious or Polygamous Alternate

Dwarf Sumac   *R. copallina*   July-Aug.   Me., Ont. & Mich., s. to Fla. & Tex.   20 (occ. 30) ft.
 (Shining Sumac)
Fragrant Sumac   *R. aromatica*   March-April   Vt., s.w. Que. & Mich., s. to Fla. & Tex.   4 (occ. 6) ft.
Poison Ivy   *R. radicans*   June-July   N.S. to Minn., s. to Fla. & Ariz.; vine, high-climbing if
                                          support is available, or forming dense ground cover
                                          in open; occ. shrub-like on tops of stumps or walls

Poison Sumac   *R. vernix*   June-July   Me. & Que. to Ont. & Minn., s. to Fla. & Tex.   20 ft.
Smooth Sumac   *R. glabra*   July-Aug.   Me. to B.C., s. to Fla. & Ariz.   10 (occ. 15) ft.
Staghorn Sumac   *R. typhina*   June-July   Me., Que., Ont. & Minn., s. to Ga., Tenn., Ill. & Ia.
 (Velvet Sumac)   20 (occ. 30) ft.

male (staminate)
flowers yellow

female (pistillate) flowers
greenish at first,
quickly turning red

fruit red,
fuzzy

Staghorn Sumac

flowers
yellowish

fruit red,
fuzzy

fruit dull; white

female
(pistillate)
flowers greenish
(see Vine section,
MP 183, for male
flowers)

Fragrant Sumac

Poison Ivy

male (staminate) flowers
yellow

female (pistillate) flowers
greenish, soon turning
red after pollinization

flowers
greenish
yellow

Smooth Sumac
(Dwarf Sumac similar)

fruit red

Poison Sumac

fruit
dull white

oison Sumac

Note: The flowers (and fruit)
of most Sumacs are terminal,
above the leaves, but those of
Poison Sumac and Poison Ivy
are along the stems, below
many of the leaves.

Smooth Sumac
(Dwarf Sumac similar)

MP
75

The three leaves at the top of this page are ½ actual size; the two leaves at the bottom are actual size.

Note wings along main leaf-stalk (rachis) of Dwarf S.; Poison S. does not have these.

Dwarf Sumac   Poison Sumac   Smooth Sumac
                                            (Staghorn Sumac similar,
                                            often very large)

Note: The end leaflet of the Poison Ivy leaf is on a definite continuation of the leaf-stalk beyond the two side leaflets. The end leaflet of Fragrant S. begins immediately above the two side leaflets.

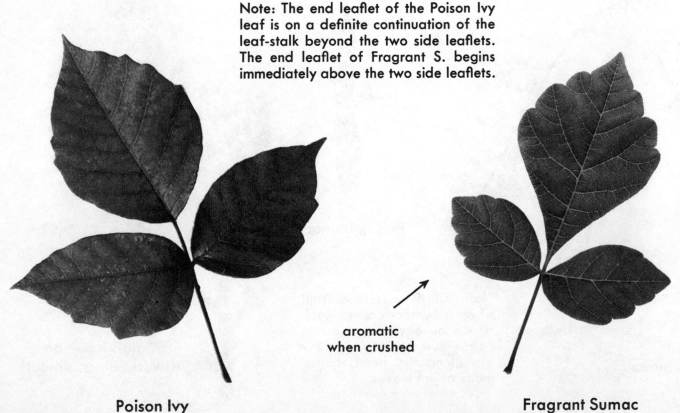

aromatic
when crushed

Poison Ivy   Fragrant Sumac

fuzzy
twig

twigs pungently aromatic when crushed;
the only Sumac with winter catkins

Staghorn
Sumac

Smooth
Sumac

Dwarf
Sumac

Poison
Sumac

Poison
Ivy

Fragrant
Sumac

barks typical of
Dwarf S.,
Fragrant S.,
Smooth S.,
Staghorn S. (only new
growth is fuzzy—see
twig, far left)

Note: Poison Ivy often climbs on trees and other objects by
means of rootlets, but
these are not always
present when the plant
is not climbing.

lenticels (dots on bark) generally
smaller than on other Sumacs

Poison Ivy

Poison Sumac

MP
77

HOLLY—*Ilex* Dioecious · Alternate

Black Alder *Ilex verticillata* (Winterberry)	June-July	Nfd. to Minn., s. to Ga. & Mo. 10-12 ft.
Inkberry *I. glabra*	June	mostly coastal, N.S. to Fla. & La.; evergreen 7-8 ft.
Large-leaved Holly *I. montana* (Mountain Winterberry; Mountain H.)	May-June	w. Mass. & N.Y., s. mostly on higher eleva- tions to S.C. & Ala. 25 ft.

female
(pistillate)
flowers

male
(staminate)
flowers

spurs

Black Alder

Large-leaved Holly

Inkberry

Note: Flowers and fruit are often on short spurs. These spurs, when present, distinguish this Holly at any time of year.

red
berries

note spur

red
berries

black
berries

MP
78

Black Alder

Large-leaved Holly

Inkberry

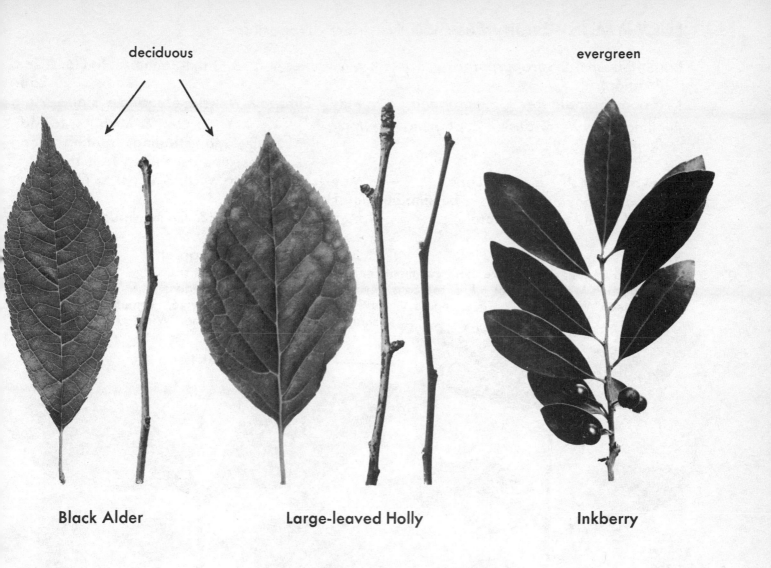

deciduous

evergreen

Black Alder   **Large-leaved Holly**   **Inkberry**

Black Alder   **Large-leaved Holly**   **Inkberry**

MP
79

EUONYMUS — *Euonymus*   usually Perfect   Opposite

Burning-bush   *E. atropurpurea*   May-June   N. Y. & Ont. to Minn., s. to Fla. & Tex.
(Wahoo)   20-25 ft.

European Spindle-tree   *E. europaea*   May   Europe & w. Asia; escaped e. states   20 ft.

Running Strawberry-bush   *E. obovata*   May   w. N. Y., s. Ont. & Mich., s. to Md.,
(Running Euonymus)   Tenn., Mo.; trailing, rooting along
stems; tips ascending to 1 ft.

Strawberry-bush   *E. americana*   May-June   N. Y. to Mich. & Ill., s. to Fla. & Tex.
(Brook Euonymus; Hearts-a-burstin; Bursting Heart)   7-8 ft.

Winged Euonymus   *E. alata*   May-June   n.e. Asia & China; much planted &
escaped   10 ft.

Wintercreeper   *E. fortunei*   June-July   China; evergreen
The two varieties below are more commonly encountered than the parent species:
   Common Wintercreeper   *E. f. radicans*   evergreen; trailing or climbing vine
   Bigleaf Wintercreeper   *E. f. vegeta*   evergreen; high-climbing vine, sometimes somewhat
shrubby; heavier than Common Wintercreeper

five petals

yellow
green

purplish
green

Strawberry-bush

Running Strawberry-bush

greenish
white

four petals

yellowis

Bigleaf Wintercreeper

Winged Euonymus

MP
80

246

four petals

purple

yellow green

Burning-bush

European Spindle-tree

outside case pink; seed-covering orange

seed-covering orange; outside case reddish pink

Burning-bush
(European Spindle-tree similar)

Strawberry-bush

outside case white; seed-covering orange

outside case reddish pink; seed-covering orange

orange red

Running Strawberry-bush

MP 81

Bigleaf Wintercreeper

Winged Euonymus

leaf hairy beneath (use lens to see clearly)

somewhat four-sided twigs

leaf widest at or above middle

four-sided twigs

leaf widest at or below middle

four-sided twigs

Burning-bush

Running Strawberry-bush

Strawberry-bush

leaves *not* hairy beneath

European Spindle-tree

evergreen

Bigleaf Wintercreeper

Common Wintercreeper

Winged Euonymus

vigorous   shoots   green with light tan stripes

small shoots green with dull orange stripes

twigs and stems with distinct wings

European Spindle-tree

Burning-bush

Strawberry-bush

Winged Euonymu

MP 82

MOUNTAIN HOLLY—*Nemopanthus mucronatus*

(not a true Holly; see MP 78)   Dioecious or Polygamous   May-June   Alternate
Nfd. to Wis. & Minn., s. to upland Va. & n. Ill.   10 ft.

female (pistillate) flowers

red berries

short, abrupt tip (mucro)

male (staminate) flowers

purple leaf-stalks

Note spurs on twigs and small branches.

BUCKTHORN — *Rhamnus* See individual species for kinds of flowers, and whether Opposite or Alternate

Alder-leaved Buckthorn *Rhamnus alnifolia*	mostly Dioecious Alternate	May-June	Nfd. to B.C., s. to n. & w. N.E., n. N.J., W. Va., n. Ia. & west 3 ft.
Common Buckthorn *R. cathartica*	mostly Dioecious mostly Opposite	May-June	Europe & Asia; naturalized here sometimes tree-like 25 ft.
European Buckthorn *R. frangula* (Alder Buckthorn)	Perfect Alternate	May-July	Europe, w. Asia & n. Africa; naturalized here 20 ft.
Lance-leaved Buckthorn *R. lanceolata*	Perfect Alternate	May	locally distributed Pa. to Neb., s. to Ala. & Tex. 6 ft.

typical Buckthorn flowers

Buckthorn flowers look alike superficially but note:

Alder-leaved B.,
   five stamens, five sepals
   and usually no petals

European B.,
   five stamens, five petals
   and five sepals

Common B. and
Lance-leaved B.,
   four stamens, four petals
   and four sepals

berries red to black

typical Buckthorn fruit

MP
84

Alder-leaved Buckthorn

European Buckthorn

Lance-leaved Buckthorn

Common Buckthorn

MP
85

Alder-leaved Buckthorn

Lance-leaved Buckthorn

side twigs often sharp-pointed

mostly opposite,
but occasionally
partly alternate

naked buds
(buds without scales)

Common Buckthorn

European Buckthorn

AMERICAN BLADDERNUT—*Staphylea trifolia*

Perfect   May   Opposite
N.H., Que. & Ont. to Minn., s. to
Ga. & Okla.   15 ft.

Seeds, when ripe,
rattle in papery pods.

ST. JOHN'S-WORT—*Hypericum*   Perfect; five petals, five sepals   Opposite

Dense-flowered St. John's-wort   *Hypericum densiflorum*   July-Sept.   L.I. to W. Va., s. Mo., s. to Fla. & Tex.   6 ft.

Shrubby St. John's-wort   *H. prolificum*   July-Sept.   N.Y., Ont. & Minn., s. to Ga. & Ark. 5 ft.

yellow flowers

Dense-flowered St. John's-wort

yellow flowers

Shrubby St. John's-wort

ST. PETER'S-WORT—*Ascyrum*   Perfect; four petals, four sepals in unequal pairs Opposite

St. Andrew's Cross   *A. hypericoides multicaule*   July-Sept.
　　　　　　Nantucket & L.I., Ky., s. Ind., s. Ill., Mo. & s.e. Kans. (for var. given here).
　　　　　　The standard species is a strictly southern shrub, more upright than this var.,
　　　　　　which has creeping stems with small branches ascending to only 6-8 in.

St. Peter's-wort   *A. stans*   July-Sept. L.I., Pa. & Ky., s. to Fla. & Tex.; upright shrub   2½-3 ft.

Note: The unequal sepals, the larger pair clasping as shown below, are distinctive of this genus.

Note markedly two-sided twigs (alternating flat sections along small stems) typical of both species.

yellow flowers

sepals removed to show seed pods

The flowers of St. Andrew's Cross are smaller, the petals narrower, than those of St. Peter's-wort, but are similar otherwise.

St. Peter's-wort　　　　　　　　　　　St. Andrew's Cross

ST. JOHN'S-WORT (con't.) typical bark and twigs

ROSE OF SHARON—*Hibiscus syriacus* Perfect July-Sept. Alternate
(Shrub Alth(a)ea)
China, India; much planted and escaped here
10 (occ. 15-20) ft.

Note: The flower of the standard type is single. There are a number of garden forms with double flowers.

The color varies from white to pink, red, blue or purple.

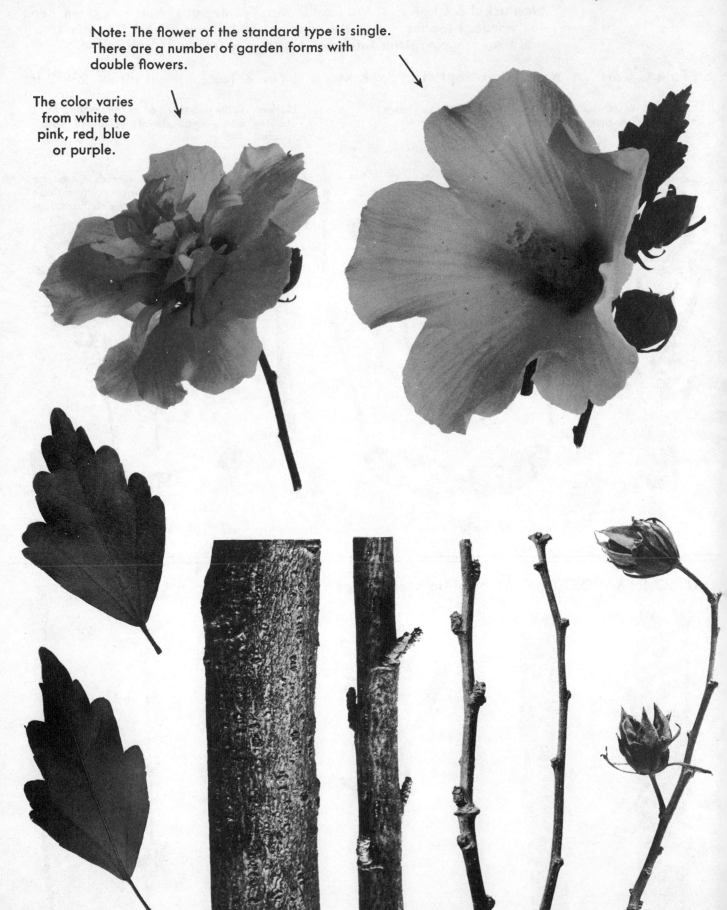

LEATHERWOOD—*Dirca palustris*
(Moosewood, Wicopy, Leatherbush)

Perfect   March-April   Alternate
N.B., Ont. & Minn., s. to Fla. & La.   6 ft.

yellow
flowers

green or
red fruit

bark fibrous, pliable and
strong; wood soft and weak

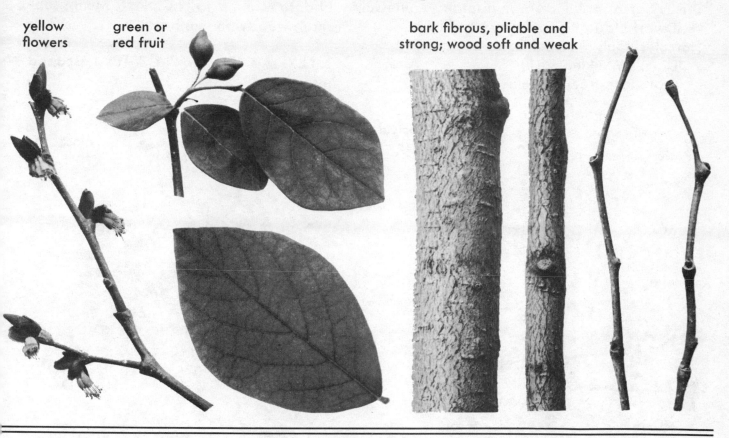

DAPHNE—*Daphne mezereum*
(Mezereum)

Perfect   March-April   Alternate
Europe; occ. escaped elsewhere   3 ft.

fragrant, pinkish-purple (occasionally white) flowers, long
before other growth starts

red
(or yellow)
berries

SARSAPARILLA or SPIKENARD — *Aralia* Perfect Alternate

Bristly Sarsaparilla *Aralia hispida* June-July Nfd. to Man., s. to N.C., Ill. & Minn.; sub-
  (Dwarf Elder; Pigeonberry; Hyeble) shrub woody only at base 3 ft.
Hercules' Club *A. spinosa*
  (Devil's Walking-stick; Aug. s. N.Y. to s. Ill., s. to Fla. & Tex.; escaped
  Toothache-tree; Angelica-tree) elsewhere 35 ft.

½ actual
size

flowers

black
fruit

Hercules' Club

MP
92

Bristly Sarsaparilla flowers similar to Hercules' Club, but smaller, less dense and branching like fruit below

black fruit

½ actual size

Bristly Sarsaparilla   (see next page for leaf of Hercules' Club)

usually, but not always, thorny

actual size

fruit remnants sometimes found in winter

Hercules' Club

Bristly Sarsaparilla

MP 93

This is all one compound leaf; these leaves are often 3 feet long.

½ actual size

Hercules' Club

BUFFALO-BERRY— *Shepherdia canadensis*   Dioecious   April-May   Opposite
(Canada Buffalo-Berry; Soapberry)   Nfd. to Alaska, s. to Me., w. Vt., n. & w. N.Y.,
n.w. Pa., n. Ohio, n.w. Ind., n.e. Ill., w. Ia. & west
6 (occ. 9) ft.

Note: Another Buffalo-berry—*S. argentea*—a western species, has silvery leaves, especially so beneath, and is often thorny, the twigs ending in sharp points (see Thorn Key, p. 1).

small yellowish flowers

fruit yellow to red

leaves slightly silvery both sides

buds and twigs somewhat silvery

typical Buffalo-berry barks

OLEASTER — *Elaeagnus*   Perfect or Polygamous   Alternate

Autumn Elaeagnus	*E. umbellata*	May-June	e. Asia; escaped elsewhere	12 ft.
Russian Olive (Oleaster)	*E. angustifolia*	June	s. Europe, w. & c. Asia; escaped else-where	20-25 ft.
Silver-berry	*E. commutata*	May-June	Canada, s. to s. Minn., S. Dak. & Utah	12 ft.

fruit red, dotted with silver or brown scales; juicy

twigs silvery or golden brown

typical Oleaster flowers

leaves silvery beneath

Autumn Elaeagnus

fruit silvery, mealy

twigs golden brown

leaves silvery both sides, brown-dotted beneath

Silver-berry

fruit yellow or silvery, mealy

twigs silvery, fuzzy

leaves dull green above, silvery beneath

Russian Olive

MP 96

Autumn Elaeagnus and Russian Olive are sometimes thorny (see also Thorn Key, p. 2); Silver-berry has no thorns.

typical Oleaster barks

SWAMP LOOSE-STRIFE—*Decodon verticillatus*
(Water Willow)

Perfect Aug.-Sept. Opposite Aquatic subshrub, woody only at base. The several-sided arching stems die back in winter except for tips, which root upon touching water or mud. In bogs and ponds, Me., s.w. Que., Ont. & Minn., s. to Fla. & La. Arching stems to 9 ft.

rose-purple flowers

leaves mostly in
whorls of three
around the stem

MP
97

DOGWOOD—Cornus Perfect Opposite (except Alternate-leaved Dogwood)

Alternate-leaved Dogwood	C. alternifolia	May-June	Nfd. to Minn., s. to Ga., Ala. & Mo.	25 ft.
Bunchberry C. canadensis (Dwarf D.; Dwarf Cornel) (see also Ground Covers, MP 169A)		May-July	Arctic, s. to Md., W. Va., Ill. & w. to Calif.; herbaceous	6-10 in.
Flowering Dogwood	C. florida	May	s. Me., s. N.H., s. Vt., N.Y., s. Ont., Ohio, Ind. & Ill., s. to Fla., Tex. & Mexico; often a tree 30 ft.	
Gray-stemmed Dogwood (Panicled D.; Gray D.)	C. racemosa	June	Me. to Ont. & Minn., s. to Del., W. Va., n. Ind., Wis. & Ia. (possibly farther south) 6 (occ. 10) ft.	
Red-Osier Dogwood (Red-stemmed D.)	C. stolonifera	May-June	Nfd. & west, s. to Md., Tenn., Ill., Ia. & Neb., s.w. to N. Mex. & Calif.	6-8 ft.
Round-leaved Dogwood	C. rugosa	May-June	Que. to Man., s. to s. N.E., Pa., W. Va. & Ia. 10 (occ. 15) ft.	
Swamp Dogwood C. amomum (Red Willow; Swamp Cornel; Silky D.)		June	Me. to Ind., s. Ga. & Ala.; western form, some- times called C. obliqua, extends range to N. Dak. & Okla.	10 ft.

Alternate-leaved
Dogwood (and other
Dogwoods; see Note, type 1)

Flowering Dogwood

Gray-stemmed Dogwood

Note: The flowers of all Dogwoods have
four petals. There are three types:
1. Alternate-leaved Dogwood. A flat
cluster of many small flowers. Red-
Osier, Round-leaved and Swamp
Dogwoods have similar flowers.
2. Gray-stemmed Dogwood. Similar to
type 1, except that the flower clusters are not flat, but more or less
conical. This shape is also evident
in fruit (see MP 100).
3. Flowering Dogwood and Bunchberry. The actual flower clusters of these two are small and relatively inconspicuous. They
are placed in the middle of four large white (or pink) bracts,
as can be seen above.

Bunchberry

MP
98

small stems often red,
especially Swamp Dogwood

**Alternate-leaved Dogwood
(Swamp Dogwood similar)**

Flowering Dogwood

large and small stems
red (there is also a yellow-
stemmed variety)

large and small stems
pinkish or greenish yellow

stems gray
(or brownish)

Red-Osier Dogwood

Round-leaved Dogwood

Gray-stemmed Dogwood

MP
99

white berries

red berries

blue-black berries on red stems

Red-Osier Dogwood

Bunchberry

blue or bluish-white berries

Flowering Dogwood

Alternate-leaved Dogwood

Swamp Dogwood (Round-leaved D. similar)

Note alternate growth (all other Dogwoods have opposite growth).

white berries

brown twigs

Note conical fruit cluster.

Gray-stemmed Dogwood

Alternate-leaved Dogwood

seven or more pairs of veins; (other Dogwoods usually six or less)

pinkish or yellowish twigs

dull red or greenish twigs; distinctive flower bud; side buds concealed

Round-leaved Dogwood

Flowering Dogwood

bright red (or maroon or green) twigs

brown pith

red leaf-stalk

Swamp Dogwood

bright red twigs

white pith

red leaf-stalk

Red-Osier Dogwood

MP 101

CLETHRA—*Clethra alnifolia* Perfect July-Sept. Alternate
(Sweet Pepperbush; Summer Sweet; White-Alder) Me. to Fla. & Tex., mostly near coast
9-10 ft.

Clethra is one of the few plants to produce side growth from new shoots.

MP
102

two types of bark, sometimes both on the same plant

MENZIESIA — *Menziesia pilosa*
(Minniebush; Allegheny Menziesia)

Perfect   May-June   Alternate
Pa. to Ga. & Ala.   6 ft.

flowers
greenish
yellow,
tipped with
red or pink

Note scattered whitish hairs on upper side and along edge of leaf
(tip down and turn to see clearly).

distinctive
whitish tip
(mucro)

Note tufts of white
hairs along mid-rib
beneath. Underside
is pale or bluish-
green.

bark rust colored;
sometimes flaky, at other times smooth

twigs hairy;
clusters of buds often at ends of twigs

MP
103

AZALEA Subgenus I
RHODODENDRON Subgenus II

—*Rhododendron*   Perfect   Alternate, but note that all produce leaves in characteristic clusters at the ends of the twigs.

Note: Azaleas and Rhododendrons, although both in the genus *Rhododendron,* are usually placed in separate subgenera. Azaleas will be treated here first, as subgenus I, and Rhododendrons separately as subgenus II, on MP 108.

Subgenus I. AZALEA Deciduous

Section 1. Rhodora: ten stamens about as long as the petals, which are divided to the base of the flower (see opposite page)

Rhodora	*R. canadense*	April-May	Nfd. & Lab., s. to N.E., N.Y. & n.e. Pa. 3 ft.

Section 2. Pentanthera: five stamens much longer than the petals, which are not divided all the way to the base of the flower

Flame Azalea	*R. calendulaceum*	May-June	s.w. Pa. to s.e. Ohio, s. to Ga. & Ala. 10 (occ. 15) ft.
Pink Azalea (Pinxterbloom; Election Pink; Wild Honeysuckle; Rose A.)	*R. nudiflorum*	April-May	s. Vt. to Ohio, s to S.C. & e. Tenn. 10 ft.
Rose Azalea (Downy Pinxterbloom; Election Pink; Honeysuckle A.; Mountain A.)	*R. roseum*	May-June	N.H. & s. Que. to Va. & Mo. 10 ft.
Swamp Azalea (Clammy A.; Swamp Honeysuckle)	*R. viscosum*	June-July	mostly in swamps, Me. to S.C., infrequently w. to Ohio & Tenn. 10 (occ. 15) ft.

There are a number of native Azaleas which in parts of their range appear to be distinct species but in other parts seem to merge into hybrid forms. As this is more confusing than important, except to the specialist, a few distinct types only are given here, with the suggestion that in general, the color, odor and form of the flowers, together with their blooming dates, are the most useful distinguishing characteristics, and that until considerable experience has been gained, identification of species should begin at flowering time. For example, Rose Azalea is almost identical in appearance with Pink Azalea. Typically, Rose Azalea blooms ten days to two weeks later than Pink Azalea, and has deeper pink, more fragrant flowers. The outside of the flower below the petals (called the tube) is sticky—a result of minute hairs ending in sticky glands (clearly visible under a lens). Pink Azalea has hairs on the tube but without the sticky glands. This constitutes an important difference, but also indicates how small the differences really are, and without the flowers it is often almost impossible to tell the two apart. In spite of these difficulties, one can always tell an Azalea from other plants at any time of year, and success with the species given here is fairly certain if attempted first at flowering time.

Subgenus II.   RHODODENDRON   Evergreen (see MP 108)

Subgenus I. AZALEA

Section 1. RHODORA: ten stamens about as long as the petals, which are divided to the base of the flower

pale to deep
rose purple

immature

mature and open

Rhodora
(April-May)

fruit typical of all Azaleas

Section 2. PENTANTHERA: five stamens much longer than the petals, which are not divided all the way to the base of the flower

five stamens (male, pollen-bearing parts) much longer than petals

apparent sixth one is the pistil (female part of flower)

petals
not divided
to base of flower
(undivided section
called the tube)

Flame Azalea
yellow to orange
red; nearly odorless
(May-June)

These three slightly smaller
but similar in form:

Pink Azalea
white (or pink);
nearly odorless
(April-May)

Rose Azalea
deep pink (or
whitish); fragrant;
tube sticky
(May-June)

Swamp Azalea
white;
fragrant
(June-July)

cluster of leaves at ends
of stems, typical of all
Azaleas (and Rhododendrons)

definite bluish cast above,
pale beneath

Rhodora

glossy, dark green

Swamp Azalea

MP
106

These seven leaves are typical of Flame, Pink and Rose Azaleas; they are extremely variable in shape, size and texture.

cluster of twigs at ends of stems, typical of Azaleas

leaf buds

flower bud

fruit

buds covered with a bloom, giving grayish or pinkish look; bud scales indistinct

flower buds

clear-cut bud scales; reddish twigs

Rhodora

Swamp Azalea

clear-cut bud scales; brown twigs

Note buds clustered at ends of twigs, typical of Azaleas.

flower bud

leaf buds

MP 107

barks typical of all Azaleas

twigs typical of Flame, Pink and Rose Azaleas

Subgenus II. RHODODENDRON

Perfect Evergreen Alternate, but with leaves in clusters at ends of twigs

Carolina Rhododendron	*R. carolinianum*	May-June	N.C.; planted & hardy north 6-7 ft.
Catawba Rhododendron (Mountain R. or Rosebay; Purple Laurel or R.)	*R. catawbiense*	May-June	W. Va. & Va., s. to Ga. & Ala. in mts.; planted & hardy north 6-7 (occ. 15) ft.
Lapland Rhododendron (Lapland Rosebay; Rock Rose)	*R. lapponicum*	June-July	Alpine summits of n. Europe, n. Asia & N. Amer.; dwarf Prostrate
Rosebay Rhododendron (Great Laurel)	*R. maximum*	June-July	n. N.E. to Ont. & Ohio, s. to Ga. & Ala.; large-growing occ. to 35 ft.

typical Rhododendron fruit

Rosebay Rhododendron
white to pink

Similar:
Carolina R.
pink, smaller
Catawba R.
lilac to purple

purple

typical Rhododendron barks

leaves pitted above; many small brown spots (scales) beneath

Lapland Rhododendron
(see also Ground Covers, MP 167)

flower buds with leafy
appendages (bracts)

leaves whitish beneath; rounded at
both ends, especially at base

leaves green or
brownish beneath;
sharply tapered at
both ends

flower buds without bracts

small leaf buds
covered by bracts

Catawba Rhododendron

leaves small;
usually rusty
beneath

Note: The leaves of
Rhododendrons are
largely clustered at
the ends of the stems
(as are those of
Azaleas; see MP 106).

MP
109

Rosebay Rhododendron

Carolina Rhododendron

LAUREL—*Kalmia* Perfect Evergreen See individual species for opposite or alternate characteristics.

Mountain Laurel *Kalmia latifolia* May-July Que. & N.B. to Ind., s. to Fla. & La.; usu-
(Spoonwood; Ivy) ally alternate 15-20 (occ. 35) ft.

Sheep Laurel *K. angustifolia* May-July Nfd. & Hud. Bay to Mich., s. to Ga.; usu-
(Lamb-kill; Wicky) ally opposite (or in 3's) 4-5 ft.

Swamp Laurel *K. polifolia* May-July mostly in bogs, Lab., Hud. Bay & west, s.
(Pale Laurel) to N.E., Pa., Ohio & Wis.; usually opposite
(or in 3's) 2-2½ ft.

white to
pink

deep pink

Mountain Laurel

Sheep Laurel

pink

Note: Mountain Laurel } flowers and fruit
Swamp Laurel } terminal (at end of stem)

Sheep Laurel — flowers and fruit
lateral (along stem
and below newest
growth)

Swamp Laurel

Mountain Laurel

Sheep Laurel

Swamp Laurel

Note distinctive flower buds, typical of Laurel.

dark, glossy green above, yellow green beneath

green above, whitish beneath; edges of leave curl under (revolute)

Mountain Laurel

Swamp Laurel

light green above, pale or whitish beneath

typical Laurel bark

Sheep Laurel

MP
112

BOG ROSEMARY—*Andromeda glaucophylla*
(Downy Bog Rosemary)

Perfect   May-July   Evergreen   Alternate
mostly in bogs, Nfd. & Lab. to Man., s. to N.E.,
N.J., W. Va., n.e. Ill. & Minn. 1½ (rarely 2½) ft.

flowers white,
tinged with pink

Note:
Swamp Laurel
on opposite
page is somewhat
similar but has
opposite leaves
with no bluish cast.

Mature leaves have a bluish
cast above and are distinctly
white below. The edges curl
under (revolute).

SAND-MYRTLE—*Leiophyllum buxifolium*

Perfect   May-June   Evergreen   mostly Alternate
Pine Barrens of N.J. & in var. s. to S.C., Tenn. & Ky.
1½ (rarely 2½) ft.

ANDROMEDA—*Pieris*   Perfect   Evergreen   Alternate (rarely Opposite)

Fetter-bush	*Pieris floribunda*	April-May	Va. & W. Va., s. to Ga. & Tenn.	6 ft.
(Mountain Andromeda)				
Japanese Andromeda	*P. japonica*	April-May	Japan; much planted here	9-10 ft.

flowers erect

Fetter-bush

flowers drooping

Japanese Andromeda

typical Andromeda bark

MP
114

fruit drooping

fruit erect

Japanese
Andromeda

flower buds

Fetter-bush

Japanese
Andromeda

flower buds

tapered
at base

rounded
at base

Fetter-bush

MP
115

LABRADOR TEA—*Ledum groenlandicum* Perfect May-June Evergreen Alternate
Greenland & Alaska, s. to N.E., n. Pa., Ohio, Mich., Wis., Minn. & west 3 ft.

Note the unusual way the fruit opens, hinging from the outer end.

leaves alternate but often in tight clusters

Stems and undersides of leaves are very woolly. This wool is at first green, later rusty orange. The leaves curl under at the edges (revolute).

LEATHERLEAF—*Chamaedaphne calyculata* Perfect April-June Evergreen Alternate
(Dwarf Cassandra)   n. Europe & n. Asia; N. Amer.: s. near coast to Ga., w. from Pa. to Ind., Wis., Minn. & B.C.
4 ft.

leaves covered with many small dots

LYONIA—*Lyonia* Perfect Alternate

Male-berry	*Lyonia ligustrina*	May-June	Me. to Mich., s. to Fla. & Tenn.; in var. w. to Okla. & Tex. 12 ft.
Stagger-bush	*L. mariana*	May-June	R.I. to Fla., w. to Tenn., Ark. & Tex. 6-8 ft.

Male-berry   Stagger-bush

LEUCOTHOE or FETTER-BUSH—*Leucothoë* Perfect Alternate

Drooping Leucothoë *Leucothoë catesbaei* April-May Va., Ga. & Tenn.; evergreen 7 ft.
(Dog Hobble; Switch-ivy)

Swamp Leucothoë *L. racemosa* May-June Mass. & N.Y., s. near coast to
(Sweet Bells; Fetter-bush) Fla., w. to Tenn. & La.; deciduous 12 ft.

deciduous

twig with all but
one leaf removed
to show
flower buds

evergreen

flower buds

flower
buds

Drooping Leucothoë

Swamp Leucothoë

MP
118

BLUEBERRY—*Vaccinium* Perfect Alternate
(Including Bilberry, Blueberry, Cranberry, Deerberry)

The Blueberries include a number of subgenera such as those shown above. For the sake of simplicity, the genus is divided here into three easily distinguishable groups:

I. BLUEBERRIES and DEERBERRY Deciduous; upright shrubs; berries blue, black, purple or green

Black Highbush Blueberry	*V. atrococcum*	May	Me. to Ont. & Mich., s. to Fla. & Ark. 10 (occ. 15) ft.
(Blue) Highbush Blueberry	*V. corymbosum*	May	Me. to Mich., s. to Fla. & La. 10 (occ. 15) ft.
Deerberry *V. stamineum* (Squaw Huckleberry)		May-June	Mass. to s.e. Mich., s. to Fla. & La. 3-5 ft.
Early Lowbush Blueberry) *V. vacillans* (Low Sweet Blueberry)		May-June	Me. to Mich., s. to Ga. & Mo. 2 (occ. 3) ft.
Late Lowbush Blueberry *V. angustifolium* (Dwarf Sweet Blueberry)		May-June	Nfd. to Sask., s. to mts. of N.H. & N.Y., w. to Minn. & in var. farther south; 1-1½ ft., prostrate on higher mts.

II. MOUNTAIN CRANBERRY and CRANBERRIES (MP 124) Evergreen; low or prostrate; berries red

Mountain Cranberry *V. vitis-idaea minus* (Rock C.; Cowberry; Lingenberry) (not a true Cranberry)		May-June	Lab. to Alaska, s. to Mass., Minn. & B.C.; creeping stems to 12 in., prostrate at higher elevations
Large Cranberry *V. macrocarpum*		June-Aug.	Nfd. to Sask., s. to N.C., Tenn., Ill., Wis. & Minn.; mostly in bogs & wet ground; creeping stems, slightly heavier than next, the tips more ascending
Small Cranberry *V. oxycoccus*		May-July	N. Amer.: s. to N.E., N.J., W.Va., Ill., Wis., Minn. & west; mostly in bogs & wet ground; stems creeping

III. BILBERRIES (MP 125) Deciduous; low or prostrate; berries blue or purple

Bog Bilberry *V. uliginosum*		May-June	n. Eu. & n. Asia; N. Amer.: s. to n. N.E., n. N.Y., Mich. & Minn., mostly in mts. (rarely to sea level); rarely to 1½ ft., usually nearly prostrate
Dwarf Bilberry *V. cespitosum*		May	N. Amer.: s. to n. N.E., n. N.Y., Mich., Minn., Colo. & n. Calif.; in alpine areas but also to sea level; rarely to 1 ft., usually in tufts only a few in. high

Like a few other genera included in this book, the Blueberries are an unstable group undergoing evolutionary changes, which makes species identification uncertain in many cases. The selection given above as representative of the main types was intentionally limited to emphasize the futility of any attempt at final definition, until such time as the group gives evidence of greater stability. The various dubious species and innumerable forms can safely be left to the dedicated hair-splitters.

MP 119

I. BLUEBERRIES and DEERBERRY

Note: Both the Blue and Black Highbush Blueberries bloom at about the same time. The leaves of the Blue Highbush are then fairly well developed, whereas those of the Black Highbush develop after the flowers.

undeveloped leaves

white flowers with some pink or red parts

(Blue) Highbush Blueberry

Black Highbush Blueberry

white, sometimes tinged with pink

Early Lowbush Blueberry

Note open, bell-shaped flowers.

white

Deerberry

white, sometimes tinged with pink

Late Lowbush Blueberry

typically smooth or slightly hairy beneath; green or yellow green above

typically woolly beneath; glossy dark green above

blue with bloom

(Blue) Highbush Blueberry
(Black Highbush B. similar, but with glossy black berries)

(Blue) Highbush Blueberry

Black Highbush Blueberry

blue with bloom

Early Lowbush Blueberry
(Late Lowbush B. similar)

(Blue) Highbush Blueberry
(Black Highbush B. similar, except as noted above)

greenish (or slightly purplish) with bloom

Deerberry
(note characteristic small leaves on fruiting branches)

leaves with distinctive bluish cast above; pale beneath

Early Lowbush Blueberry

Late Lowbush Blueberry

Deerberry

MP 121

I. BLUEBERRIES and DEERBERRY (con't.)

twigs reddish or green;
smooth; not dotted

twigs reddish or green;
larger ones, especially,
covered with many minute dots

twiggy look,
typical of all
Blueberries

twigs numerous;
mostly very fine

Early Lowbush Blueberry

Late Lowbush Blueberry

reddish or green

reddish, usually
with some bloom

Highbush Blueberry
(both Black and Blue)

Deerberry

MP
122

288

Blueberries have extremely variable but distinctive barks. The Huckleberries, which are often confused with Blueberries, have uniform gray bark (see MP 127).

The older stems of Blueberries have rough, flaky bark; the young growth is smooth: maroon, red, green, yellow or brown, often all on the same plant.

Lowbush Blueberry (both Early and Late)

Highbush Blueberry (both Black and Blue)
(Deerberry similar)

II. MOUNTAIN CRANBERRY and CRANBERRIES   Evergreen; all have red berries, sometimes white before maturing

leaves and fruit
smaller at higher
elevations

Mountain Cranberry

Large Cranberry

leaves rounded
at tips;
mostly flat;
pale beneath

MP
124

Small Cranberry

leaves pointed
at tips;
curl over at
edges (revolute);
white beneath

III. BILBERRIES Deciduous; berries blue or purple with bloom

typical
Bilberry
flowers
white,
tinged
with pink

typical Bilberry fruit

Bog
Bilberry

leaves leathery;
bluish green;
twigs stout

leaves not leathery;
yellow green;
twigs slender

Dwarf
Bilberry

COMMON MATRIMONY-VINE—*Lycium halimifolium*

Note: The Chinese Matrimony-vine—*L. chinense*—is similar but usually thornless and has wider leaves.

Perfect   May-Sept.   Alternate
s.e. Europe & Asia; escaped elsewhere; rambling shrub, sometimes supported on trellises or against buildings   10 ft.

flowers purple the
first day, quickly
turning tan

red fruit

usually
sparingly
thorny

HUCKLEBERRY—*Gaylussacia* Perfect Alternate

Black Huckleberry G. baccata (Highbush Huckleberry)	May-June	Nfd. to Man. & e. Minn., s. to Ga. & La.	3 ft.
Blue Huckleberry G. frondosa (Dangleberry, Tangleberry)	May-June	N.H. to La. near coast	6-7 ft.
Dwarf Huckleberry G. dumosa (Swamp Huckleberry)	May-June	Nfd. to La. near coast, infrequent inland to W. Va. & Tenn.	1½ ft.
Evergreen Huckleberry G. brachycera (Box-Huckleberry)	April-May	Del. & Pa. to Tenn.	1½ ft.

flowers mostly reddish

Note: Both Blue and Dwarf Huckleberries have leaf-like bracts on the flower stems; bracts of Dwarf Huckleberry persist into fruit.

flowers white, tinged with pink

flowers white, tinged with pink or red

true leaf

bracts

Black Huckleberry   **Blue Huckleberry**   **Dwarf Huckleberry**   **Evergreen Huckleberry**

Note: The berries of all Huckleberries contain ten fairly large seeds. (Blueberries have many small seeds.)

shiny black berries

berry cut in half, showing ten seeds

blue berries with bloom

shiny black berries; note leaf-like bracts

blue berries with bloom

Black Huckleberry   **Blue Huckleberry**   **Dwarf Huckleberry**   **Evergreen Huckleberr**

Note: The deciduous Huckleberries have yellow resin dots on the leaves (at least, always on underside). These are difficult to see without a magnifying glass, but are very useful in case of doubt.

green or yellow green above

somewhat bluish green above

dark green and glossy above, with distinct bristle tip (mucro)

pale beneath

Black Huckleberry

Blue Huckleberry

Dwarf Huckleberry

Evergreen Huckleberry

leaf buds with two to three exposed scales (use lens)

twigs reddish, without (or with slight) bloom

twigs pinkish, with heavy bloom

leaf buds with four to five exposed scales; twigs reddish, without (or with slight) bloom

flower bud

flower bud

leaf buds (small ones)

leaf buds (small ones)

flower buds

Black Huckleberry

Blue Huckleberry

Dwarf Huckleberry

typical Huckleberry bark (fairly uniform throughout)

MP 127

SILVERBELL—*Halesia carolina*   Perfect   April-May   Alternate
(Snowdrop Tree; Opossum-wood; Shittimwood)   W. Va. to Ill., s. to Fla. & Tex. 35 ft.

FRINGE-TREE—*Chionanthus virginicus*
Dioecious (or functionally so) May-June Opposite (occ. partly Alternate)
N.J., e. Pa. & W. Va., s. to Fla. & Tex. 30 ft.

dark blue

mostly opposite,
but occasionally
partly alternate

MP
129

COMMON LILAC — *Syringa vulgaris* Perfect April-May Opposite
(Old-fashioned Lilac)

There are many Lilac species and hybrids, all originating in s.e. Europe & Asia, but this is the only one frequently escaping in this country. 20 ft.

flowers purple before opening, pale lavender when open and very fragrant (also a white variety)

bark shaggy when larger

FORSYTHIA—*Forsythia*   Perfect   April-May   Opposite
(Golden-bells)

China   10 ft.

There are a number of species and hybrids. The following are the standard types:

all with
yellow flowers

Weeping Forsythia
F. suspensa

pendulous branches; hollow stems except solid at nodes (joints); two kinds of leaves (see below)

Greenstem Forsythia
F. viridissima

upright branches; stems with chambered pith; simple leaves only (bottom type)

Border Forsythia
F. intermedia

a cross between the above two

PRIVET—*Ligustrum*   Perfect   Opposite

California Privet	*L. ovalifolium*	July	Japan; escaped elsewhere	15 ft.
Common Privet	*L. vulgare*	June	Europe & N. Africa; escaped elsewhere	15 ft.
Ibota Privet	*L. obtusifolium**	June	Japan; escaped elsewhere	10 ft.

(Regal Privet—*L. o. regelianum*—is a dwarf variety, not over 3-4 ft.)

*The name Ibota is commonly used for this Privet although more correctly applied to another species—*L. ibota*.

short petals (less than half total length of individual flower); flower clusters small, produced all along stems, not just at ends

long petals (at least half total length of individual flower); flower clusters large, mostly only at ends of branches

Common Privet

Note: Ibota and Common P. bloom at about the same time. California P. blooms two to three weeks later. It has short petals like Ibota P., but the flower clusters are large and at the ends of the stems like those of Common P.

Ibota Privet

fruit clusters along small branches as well as at the ends

fruit clusters mostly at ends of branches

glossy black

black with bloom, giving bluish or purplish cast to berries

black with bloom, giving bluish cast to berries

Ibota Privet   **California Privet**   **Common Privet**

California Privet

Common Privet

Ibota Privet

California Privet

Common Privet

Ibota Privet

ELDER or ELDERBERRY—*Sambucus*   Perfect   Opposite

Common Elder *Sambucus canadensis* (American Elder)	June-July	N.S. to Man., s. to Fla. & Tex.	12 ft.
Red-berried Elder *S. pubens* (Scarlet Elder)	April-May	Nfd. to Alaska, s. to Ga., Tenn., Ind., Ia. & west	25 ft.

flowers and fruit
in flat clusters

dark purple or
black berries

red berries

Common Elder

Red-berried Elder

typical
Elder leaf

Common Elder:
green leaf-stalk

Red-berried Elder:
purplish leaf-stalk

white pith

brown pith

Common Elder

Red-berried Elder

MP
135

VIBURNUM— *Viburnum* Perfect   Opposite

Arrow-wood *V. dentatum*	May-June	N.B. to Mich., s. to Ga. & Tenn.	15 ft.
Black-haw *V. prunifolium* (Stag-bush, Sweet-haw)	April-May	Conn. to Mich., s. to Fla. & Tex.	15 ft.
Downy Viburnum *V. rafinesquianum* (Downy Arrow-wood)	May-June	N.H. & Que. to Minn., s. to Ga. & Ark.	6 ft.
Highbush Cranberry *V. trilobum* (Pimbina)	May-July	N.B. to B.C., s. to N.E., Pa., W. Va., Ill., Ia. & west	12 ft.
Hobble-bush *V. alnifolium* (Witch-hobble; Tangle-legs; Moosewood)	April-June	N.B. to Mich., s. to Ga. & Tenn. stems often arching & rooting at tips	10 ft.
Maple-leaved Viburnum *V. acerifolium* (Maple-leaved Arrow-wood; Dockmackie)	May-June	N.B. to Mich. & Wis., s. to Ga. & Tenn.	6 ft.
Nanny-berry *V. lentago* (Sheep-berry; Sweet Viburnum; Wild Raisin)	May-June	Me. & Que. to Man., s. to Ga., Ill. & Colo.	30 ft.
Smooth Withe-rod *V. nudum* (Possum-haw; Swamp-haw)	June-July	Conn. to e. Tex. & Fla.	15 ft.
Southern Arrow-wood *V. pubescens*	June-July	Mass. to Pa., s. to Fla. & Tex.	10 ft.
Withe-rod *V. cassinoides*	June-July	Nfd. to Ont., Mich. & Wis., s. to N.C. & Ala.	10 ft.

Note: There are two main types of flowers:

I. with fertile flowers only

II. with large marginal, sterile flowers, and small fertile ones forming the central flower cluster

The flowers of all Viburnums have five petals.

Type I
Arrow-wood
Black-haw
Downy Viburnum
Maple-leaved Viburnum
Nanny-berry
Smooth Withe-rod
Southern Arrow-wood
Withe-rod

Type II
Highbush Cranberry
Hobble-bush

MP
136

red (not always so large)

red, eventually
turning dark
purple or black

yellow to
pink, finally
dark blue
with bloom

Highbush Cranberry

Hobble-bush

dark blue

pink, turning dark blue,
but with a heavy bloom
giving berries light blue
appearance

Arrow-wood
(Southern A. similar)

Nanny-berry
(Black-haw similar)

glossy black,
sometimes with a
bloom giving berries
bluish look

Withe-rod
(Smooth W. similar)

sometimes more oval
or flattened, like
Maple-leaved Viburnum at
extreme left

considerable variation
in size and shape,
ome berries being flattish,
others round

Maple-leaved Viburnum
(Downy V. similar)

MP
137

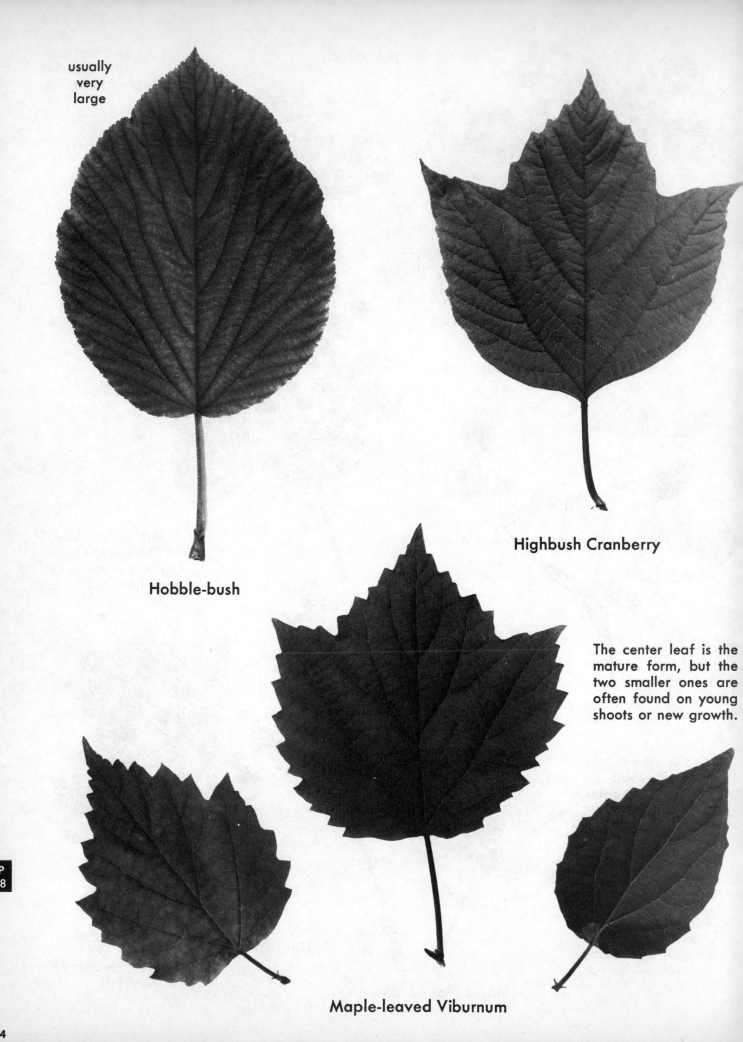

usually
very
large

Hobble-bush

Highbush Cranberry

The center leaf is the
mature form, but the
two smaller ones are
often found on young
shoots or new growth.

Maple-leaved Viburnum

MP
138

Note leaves with irregular teeth, or wavy-edged.

← wavy-edged leaf-stalk

Nanny-berry

reddish leaf-stalk →

Black-haw

Withe-rod

Arrow-wood

Southern Arrow-wood

Smooth Withe-rod

fuzzy beneath, velvety to touch →

Note stipules (leafy appendages to leaf-stalk). →

short leaf-stalks (especially those of outermost pair of leaves on any twig)

Downy Viburnum

naked buds (no scales covering buds)

flower bud

buds usually with six scales (sometimes four)

brown buds

buds with four scales

Hobble-bush

Downy Viburnum

Southern Arrow-wood

greenish buds
(sometimes tinged with red)
with two scales

brown buds
with four scales

twigs often several-sided

twigs round

Highbush Cranberry

Arrow-wood

Maple-leaved Viburnum

The buds on this page all have two scales.

flower buds   growth buds   buds rust colored

side buds
long

Withe-rod

side buds
short

Smooth Withe-rod

buds light gray or pinkish;
flower buds long-pointed

side buds
long

Nanny-berry

buds usually reddish
(sometimes with slight bloom);
flower buds short-pointed

side buds
short

Black-haw

MP
141

Nanny-berry

Hobble-bush

Black-haw

Highbush Cranberry

Arrow-wood

Downy Viburnum

Withe-rod

(Smooth Withe-rod and
Southern Arrow-wood similar)

Maple-leaved Viburnum

BUTTONBUSH—*Cephalanthus occidentalis* Perfect July-Sept. Opposite
(Honeyballs; Globeflower) usually in swamps or edges of ponds or streams;
N.B. to Ont. & Minn., s. to Fla., Okla. & Tex., w. to Mex. & Calif. 15 ft.

leaves and buds opposite,
frequently in whorls of threes
or fours around the stems

MP
144

CORALBERRY and SNOWBERRY—*Symphoricarpos*   Perfect   Opposite

Coralberry   *S. orbiculatus*   July-Aug.   Pa. to Minn., S. Dak. & Colo., s. to Fla. & Tex.;
   (Indian Currant; Buckbush)   as an escape, n. to N.E. & N.Y.   6 ft.
Snowberry   *S. albus*   May-Sept.   N.S. to Alb., s. to w. Mass., n. N.Y., Mich. &
   Minn.; as an escape, s. to Va.   4 ft.

flowers pinkish purple

fruit maroon or pinkish purple

flowers pink

fruit white

stems with pith   Coralberry   stems hollow   Snowberry

MP
145

311

HONEYSUCKLE—*Lonicera* Perfect Opposite

I. VINES: all have hollow stems

Hairy Honeysuckle	*Lonicera hirsuta*	May-July	Que. to Sask., s. to w. Vt., N.Y., Pa., Ohio, Mich., Minn. & Neb.; high-climbing
Japanese Honeysuckle	*L. japonica*	June-Sept.	Asia; commonly escaped; semi-evergreen; trailing or high-climbing
Limber Honeysuckle (Glaucous H.; Smooth-leaved H.; Mountain H.)	*L. dioica*	May-June	Me. & Que. to Sask., s. to Ga., e. Tenn., n.e. Ill. & Ia.; in var. to Kans.; bushy or slightly climbing
Trumpet Honeysuckle	*L. sempervirens*	May-Sept.	Me. to Mich. & Ia., s. to Fla. & Tex.; semi-evergreen; high-climbing

II. UPRIGHT SHRUBS (MP 150)

A. Stems with solid white pith; native

American Fly-Honeysuckle (Twinberry)	*L. canadensis*	April-May	Me. & Que. to Sask., s. to s. N.E., Pa., Tenn. to Minn. 5 ft.
Mountain Fly-Honeysuckle (Blue Honeysuckle)	*L. villosa*	May-July	Lab. to Man., s. to n. N.E., Pa., Mich. and Minn. Low (rarely 3 ft.)
Swamp Fly-Honeysuckle	*L. oblongifolia*	May-June	N.B. to Man., s. to n. N.E., N.Y., Pa., Mich. & Minn. 5 ft.

B. Hollow stems; foreign but escaping

Amur Honeysuckle	*L. maackii*	May-June	e. Asia	15 ft.
European Fly-Honeysuckle	*L. xylosteum*	May-June	Europe	10 ft.
Morrow Honeysuckle	*L. morrowii*	May-June	Eurasia	8 ft.
Ta(r)tarian Honeysuckle	*L. tatarica*	May-June	Eurasia	10 ft.

I. HONEYSUCKLE: VINES

coral-red flowers; red berries

regular, evenly divided flowers

unevenly divided (two-lipped) flowers

flowers pale yellow (sometimes turning light orange)

Trumpet Honeysuckle

black berries

Note: Japanese Honeysuckle blooms and fruits all along the stems; other Honeysuckle vines bloom and fruit only at the ends of the stems.

Limber Honeysuckle
(Hairy H. similar, but slightly larger flowers, very deep yellow)

flowers white, turning yellow

unevenly divided (two-lipped) flowers

red berries

Japanese Honeysuckle

Limber Honeysuckle
(Hairy H. similar)

I. HONEYSUCKLE: VINES (con't.)

The Honeysuckles on this page are semi-evergreen; those on the next page are deciduous.

Note connate leaf (two leaves joined together). The upper pairs of leaves of most Honeysuckle vines are connate. The Japanese Honeysuckle (shown at right) is one of the rare exceptions.

upper leaves distinct, *not* connate

cut-leaf variety occasionally encountered

Japanese Honeysuckle

upper pairs of leaves connate

leaves white beneath

Trumpet Honeysuckle
(high-climbing)

barks typical of Honeysuckle vines
(all have hollow stems)

MP
148

upper pairs of
leaves connate

leaves white
beneath

Limber Honeysuckle
(bushy or slightly climbing)

upper leaves connate; leaves
hairy; hairs especially evident
along the edges; veins prominent

leaves green
beneath

Hairy Honeysuckle
(high-climbing)

II. HONEYSUCKLE: UPRIGHT SHRUBS   A. Solid Stems

individual flowers regular
(petals all about equal)

individual flowers
unevenly split
(two-lipped)

pale
yellow

yellowish
white

pale
yellow

Mt. Fly-Honeysuckle

Swamp Fly-Honeysuckle

American Fly-Honeysuckle

B. Hollow Stems

pink or
white
(see below)

Tartarian Honeysuckle
flowers pink, or if white,
not turning yellow

(Morrow H. similar,
but with white flowers
turning yellow)

white,
turning
yellow

pale
yellow

Amur Honeysuckle

**European
Fly-Honeysuckle**

A. Solid Stems

red

blue with bloom

red

American Fly-Honeysuckle

Mountain Fly-Honeysuckle

Swamp Fly-Honeysuckle

B. Hollow Stems

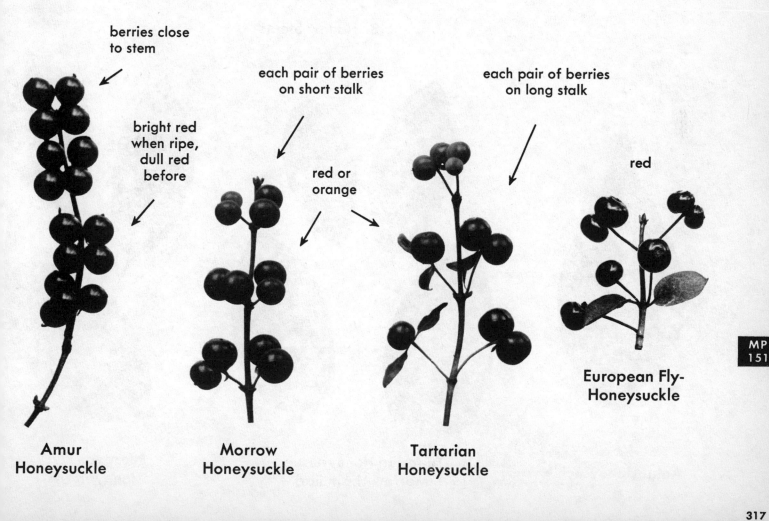

berries close
to stem

each pair of berries
on short stalk

each pair of berries
on long stalk

bright red
when ripe,
dull red
before

red or
orange

red

**Amur
Honeysuckle**

**Morrow
Honeysuckle**

**Tartarian
Honeysuckle**

**European Fly-
Honeysuckle**

MP
151

A. Solid Stems

Note reddish leaf-stalks.

Mountain Fly-Honeysuckle

American Fly-Honeysuckle

Swamp Fly-Honeysuckle

B. Hollow Stems

Amur Honeysuckle

Tartarian Honeysuckle
(Morrow H. similar)

European Fly-
Honeysuckle

MP
152

A. Solid Stems

B. Hollow Stems

American Fly-Honeysuckle **Mountain Fly-Honeysuckle** **Swamp Fly-Honeysuckle**

Amur, Morrow and Tartarian Honeysuckles similar **European Fly-Honeysuckle**

typical Honeysuckle barks (both solid and hollow stems)

barks typical of all Honeysuckles included here (except Amur, shown at right)

Amur Honeysuckle

MP 153

BUSH-HONEYSUCKLE – *Diervilla lonicera*   Perfect   May-July   Opposite

Nfd. to Sask., s. to N.C., e. Tenn., n.e. Ill. & n.e. Ia.   3 ft.

Note: This is not a true Honeysuckle; although the flowers, bark and twigs are somewhat similar, the fruit of true Honeysuckles is a berry and their leaves do not have teeth (see MP 150-153).

yellow flowers

GROUNDSEL-BUSH — *Baccharis halimifolia*
(Consumption-weed; Sea Myrtle)

Dioecious   Aug.-Sept.   Alternate
coastal marshes, Mass. to Fla., Tex. & Mex.
12 ft.

female (pistillate) flowers

cottony fruit (pappus)

male (staminate) flowers

green twigs

MARSH-ELDER — *Iva frutescens oraria*   (see flower note below) Aug.-Sept.
mostly Opposite

The species *Iva frutescens* is a southern plant: Va. to Fla. & Tex. The variety given here is found in salt marshes from s. N.H. to Va. It is a subshrub growing to 3 ft. but dying back in winter.

Note: Both male (staminate) and female (pistillate) flowers develop in each individual flower cluster.

fruit

lower leaves opposite; upper ones often alternate

Note: Mostly opposite, but the upper parts often have alternate characteristics.

NEEDLE-LEAVED SHRUBS
(all evergreen)

NEEDLE-LEAVED SHRUBS — *all evergreen*

The plants in this section include all Needle-leaved Shrubs, whether large or small, upright or prostrate. They comprise only ten genera and are shown on the following four pages. As there are so few plants, no Key is necessary, and both genus and species identification is accomplished by turning directly to this section whenever a Needle-leaved Shrub is in question. A quick perusal will indicate what is meant by the term "needle-leaved," and with the exception of the three plants below, which actually are not needle-leaved, there should be no confusion concerning which plants belong in this section.

The shrubs below will be found on the pages indicated for the Broad-leaved Upright Shrub section. Bog Rosemary, a close relative of Laurel, has very narrow leaves but they are not needle-like. Gorse produces no actual leaves, only leaf-stalks that are extremely sharp and thorn-like. Scotch Broom, although normally producing small tri-foliate leaves, sometimes is found with almost bare stems in summer, or with only a few undeveloped leaves as shown below.

Bog Rosemary 113A   Gorse 67B   Scotch Broom 65B

AMERICAN YEW—*Taxus canadensis*
(Ground Hemlock; Canada Yew)

Dioecious, rarely Monoecious   March
Nfd. to Man., s. to N.E., w. Va., W. Va.,
n. Ky., n. Ill., n.e. Ia. & Minn.   3 (occ. 6) ft.

inconspicuous flowers   conspicuous red fruit

Note: Leaves are yellow green beneath. (Hemlock, a tree, has somewhat similar-appearing needles, but they are whitish beneath.)

JUNIPER—*Juniperus*

Dioecious or Monoecious   Flowers inconspicuous

Common Juniper *Juniperus communis*
(Ground Juniper—*J. c. depressa*—is prostrate)

Europe & Asia; N. Amer.: s. to mts. of Ga., w. to Ill., Ia., & Calif.   6 ft. (rarely a tree, to 35 ft.)

Creeping Juniper *J. horizontalis*
(Creeping Savin)

N.S. to Alb., s. to n. N.E., n. N.Y., Mich. & west   Prostrate

fruit blue with bloom

leaves all
needle-like

leaves all
scale-like

Note: Red Cedar—*J. virginiana*—is a tree that has leaves both needle-like and scale-like on the same plant.

Common
Juniper

Creeping
Juniper

BROOM CROWBERRY—*Corema conradii*

Dioecious   April-May
    rare: Nfd. near coast to Mass., Shawangunk Mts. of
    N.Y., & N.J. pine barrens   18-24 in.

male (staminate)
flowers purple
brown to brown

female (pistillate)
flowers brown

fruit minute and
inconspicuous

CROWBERRY—*Empetrum nigrum*

(Curleberry)   Dioecious or Monoecious   April-May
circumpolar, s. to Me. coast, alpine areas of N.E., N.Y.,
one isolated spot on e. L.I., n. Mich., n. Minn. & n. Calif.
                                    Prostrate

flowers inconspicuous,
along stems and lost
in among the leaves

fruit black
or purple

HEATH—*Erica*   Perfect   Europe; naturalized e. U.S.

There are a number of species, two of the most common
being:

Cornish Heath   *Erica vagans*   July-Oct.   1½-2 ft.
Spring Heath   *E. carnea*   Feb.-May   1½-2 ft.

Most Heaths are late-bloomers,
but Spring Heath blooms in late
winter and early spring. Heath
flowers vary from rosy red or
pinkish lavender to white, rarely
yellow.

HEATHER—*Calluna vulgaris*

Perfect   July-Sept.
Europe, e. Asia; naturalized e. U.S.
    1½ (rarely 3) ft.

flowers lavender
to pink

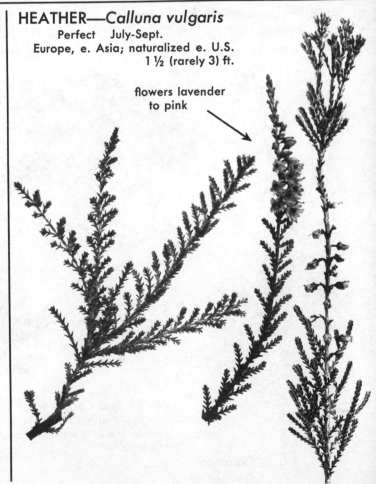

MOUNTAIN HEATH—*Phyllodoce coerulea*
Perfect   June-Aug.
circumpolar, s. to alpine areas of Me. & N.H.   6 in.

flowers light lavender
to purple

CASSIOPE—*Cassiope hypnoides* Perfect   June-July
circumpolar, s. to alpine areas of Me. & N.H. Prostrate; tufts

flowers white
on red stems
with red calyx
(star at base
of flower)

BEACH HEATHER—*Hudsonia*   Perfect

Golden Heather	H. ericoides	May-June	Nfd. & N.B., s. to N.H. & interruptedly s. to N.J., Va. & N.C. 12 in.
Woolly Heather	H. tomentosa	May-June	chiefly near coast, N.B. to N.C., also near Great Lakes; in var. farther inland 12 in.

Note: The individual needles of Woolly Heather are much closer to the stems and less clear-cut than those of Golden Heather. Flowers and fruit of Woolly Heather have little or no stalks (sessile), whereas those of Golden Heather are on distinct stalks.

Both Beach Heathers have yellow flowers.

Note distinct
flower stalks.

Note that fruit is on
distinct stalks.

Woolly Heather

Golden Heather

CLUBMOSS—Lycopodium
(Ground Pine)

This genus does not have true flowers; reproduction is chiefly by means of spores. The Clubmosses are found in most of the range covered here.

Ground Cedar—*Lycopodium complanatum*   fruits July-Sept.
(Creeping Jenny; Running or Trailing Evergreen; Ground Pine)
Note: Another Clubmoss, *L. tristachyum*, has many of the same common names as *L. complanatum*, and except for being more deeply rooted, is almost identical with it.

Running Clubmoss—*L. clavatum*   fruits July-Sept.
(Coral or Staghorn Evergreen; Buckhorn; Wolf's-claws)

Shining Clubmoss—*L. lucidulum*   fruits July-Sept.
Note: The fruiting spores are along the stems and not terminal, as with the other Clubmosses shown here.

Tree Clubmoss—*L. obscurum*   fruits July-Nov.
(Flat-branch Ground Pine)

This page all ½ actual size

fruiting strobile, producing pollen-like spores

spores along stems (no strobiles)

fruiting strobiles

Shining Clubmoss

Tree Clubmoss

Running Clubmoss

fruiting strobiles

Ground Cedar

BROAD-LEAVED GROUND COVERS

The plants shown on the following eight pages are all small, broad-leaved plants that normally do not grow over 12 inches in height (mostly less). Some of them cover a considerable area of ground, forming dense mats. Some are tufted small plants; others have fairly long runners, but do not have heavy woody stems. The distinction between a trailing ground cover and a trailing vine is purely arbitrary, and size has been the deciding factor in this case. A perusal of both sections will make this point clear.

Some ground covers belong to genera that include upright shrubs. In such a case, a cross reference is made to the Master Page giving both types together. In all other cases, the ground covers are shown here only.

Sand Myrtle, below, is not a ground cover, but a small upright shrub; it is shown here as it is sometimes quite low and might be confused with one of the small evergreen ground covers. It will be found in the Upright Shrub section, as indicated below.

Sand Myrtle 113B

COMMON BEARBERRY—*Arctostaphylos uva-ursi*
(Kinnikinic; Mealberry)

Perfect   April-June   mostly Alternate
N. Amer.: s. to Va., Ill. & west
Prostrate; creeping

flowers white,
tinged with pink

fruit red

PACHISTIMA—*Pachistima canbyi*
(Mountain Lover; Cliff Green; Rat Stripper)

Perfect   April-May   Opposite
mts. of Va. and W. Va.; sparingly in Ky. & Ohio
Matted, 10-12 in.

flowers greenish brown

fruit an
inconspicuous
capsule

MP
164

GC

330

CRANBERRY—*Vaccinium*   (see MP 119)   Perfect   Alternate

Large Cranberry—*V. macrocarpum*   June-Aug.   Nfd. to Sask., s. to N.C., Tenn., Ill., Wis. & Minn. Creeping
Small Cranberry—*V. oxycoccus*   June-July   N. Amer.: s. to N.E., N.J., W. Va., Ill., Wis., Minn. & west
                                                                                                        Creeping

leaves largely flat;
tips rounded

leaves curl over at
edges; tips pointed

red

red

red

Large Cranberry   **Small Cranberry**

VERGREEN HUCKLEBERRY
(Box-Huckleberry)   *Gaylussacia brachycera*
(see MP 126)   Perfect   April-June   Del. & Pa. to Tenn.
Creeping, with ascending stems to 18 in.

flowers white, tinged
with pink and red

fruit blue
with bloom

FLOWERING MOSS—*Pxyidanthera barbulata*
(Pyxie)   Perfect   April-May
N.J. to N.C. in sandy barrens   Prostrate tufts

flowers white;
fruit inconspicuous

MP
165

GC

BEARBERRY WILLOW—*Salix uva-ursi*

(see MP 1)   Dioecious   June-July

Lab. to Alaska, s. to mts. of n. N.E. & n. N.Y.

Prostrate; matted

reddish female (pistillate) flowers

grayish or yellow male (staminate) flowers

ALPINE BEARBERRY—*Arctous alpinus*

Perfect   April-June

Arctic, s. to mts. of Me. & N.H.   Prostrate; matted

black or purplish berries

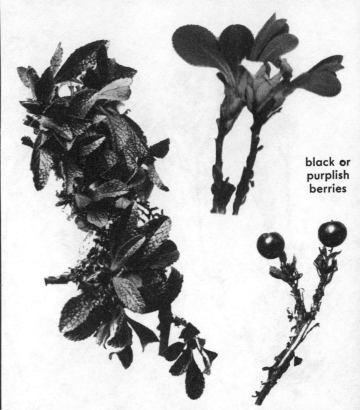

THREE-TOOTHED CINQUEFOIL

Potentilla tridentata   (see MP 64)

Perfect   May-Oct.   Arctic, s. to N.E. & N.Y., w. to N. Dak., s. in mts. to Ga.; woody only at base

Low; matted

white flowers

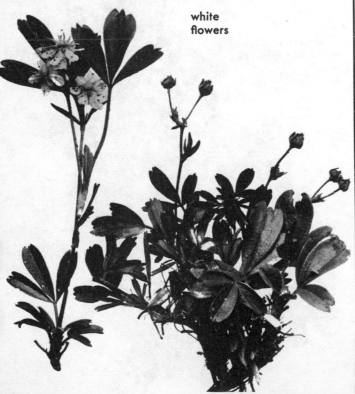

BILBERRY—*Vaccinium*   (see MP 119)   Perfect

Bog Bilberry   *V. uliginosum* May-June   n. N. Amer., s.

(Alpine B.)   to N.E., n. N.Y., Mich. & Minn.   Low; prostrate

Dwarf Bilberry   *V. cespitosum* May   n. N. Amer. s. to N.E., n. N.Y., Mich., Minn. & west   Low; tufts

leaves thin, green

white to deep pink

Dwarf Bilberry

typical Bilberry flowers

blue or purple with bloom

leaves leathery; bluish green

Bog Bilberry

typical Bilberry fruit

MP 166

GC

LAPLAND RHODODENDRON
Rhododendron lapponicum   (see MP 104)
Perfect   June-July   Arctic, s. to mts. of Me. N.H. &
N.Y.; also in Wis.   Prostrate; matted

flowers purple
or magenta

leaves pitted above,
many small brown
spots (scales) beneath

MOUNTAIN CRANBERRY
Vaccinium vitis-idaea minus   (see MP 119)
Perfect   May-June   Arctic, s. to N.E., n. Minn. & west
Matted, 6-12 in.; prostrate at higher elevations

flowers white,
tinged with
pink

berries red,
usually small
at high altitudes,
larger elsewhere

ALPINE AZALEA—*Loiseleuria procumbens*
Perfect   June-Aug.   Arctic, s. to mts. of Me. & N.H.
                   Prostrate; matted

leaves with
definite groove
on upper side

fruit red at first,
brown when mature

pink flowers

DIAPENSIA—*Diapensia lapponica*
Perfect   June-July   Arctic, s. to mts. of N.H. & N.Y.
                   Prostrate; tufts

fruit

leaves without
definite groove
on upper side

white flowers
with very yellow
stamens

MP
167

GC

PARTRIDGE-BERRY—*Mitchella repens*   Perfect   June-Aug.   Opposite
N.S. to Ont. & Minn., s. to Fla., Ark. & Tex.   Trailing

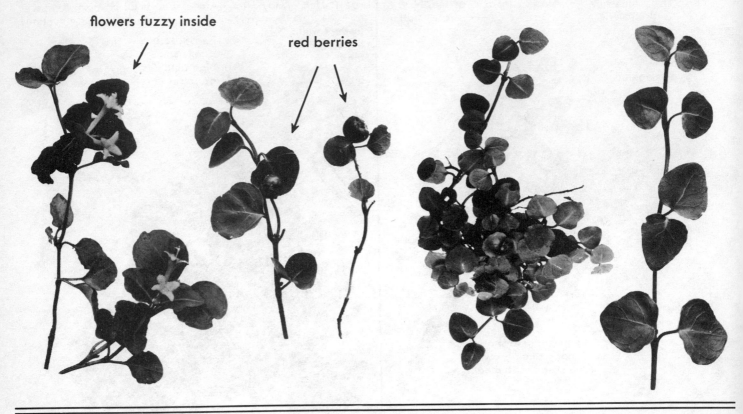

flowers fuzzy inside

red berries

TWINFLOWER—*Linnaea borealis americana*
Perfect   June-Aug.   Opposite
Lab. to Alaska, s. to L.I., W. Va., Wis. & west   Trailing

fruit fuzzy

flowers pink, red inside

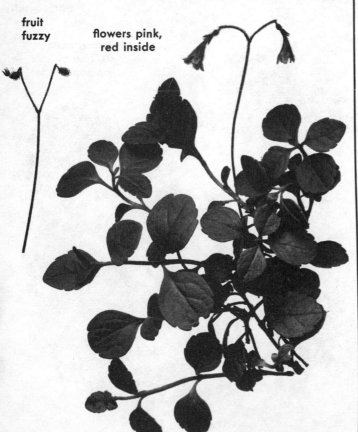

CREEPING SNOWBERRY—*Chiogenes hispidula*
Perfect   May-June   Alternate
Nfd. to B.C., s. to N.C., W. Va., Wis., Minn. & west
Trailing

flowers inconspicuous, on underside of stems

Leaves and fruit have wintergreen odor and taste when crushed.

white berries

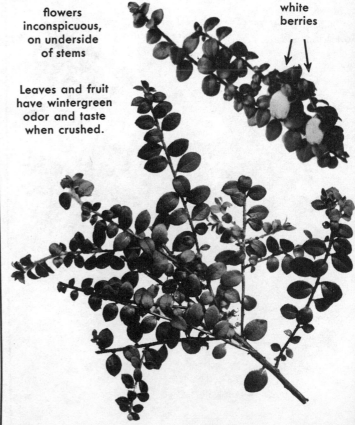

MP 168

GC

BUNCHBERRY—*Cornus canadensis* (see MP 98) Perfect May–June Deciduous Opposite
(Dwarf Cornel; Dwarf Dogwood) (leaves in whorls at ends of stems)
Arctic, s. to Md., W. Va., Ill. & west Upright, 6-10 in.

red
berries

TRAILING ARBUTUS—*Epigaea repens* Perfect April-May Evergreen Alternate
(Mayflower) Lab. to Sask., s. to Ga., Tenn., Ind., Wis. & Minn.
Prostrate; trailing

flowers white,
tinged with
pink; very
fragrant

Covering opens,
when mature, to
expose seeds on
surface of fruit,
like a strawberry.

MP
169

GC

WINTERGREEN—*Gaultheria procumbens*
(Checkerberry; Teaberry)

Perfect   May-Sept.   Nfd. to Man., s. to Ga. & Ala. Upright plants to 6 in., with underground stems, often covering considerable area

Leaves and fruit have wintergreen odor and taste when crushed.

fruit red

PIPSISSEWA—*Chimaphila*   Perfect   Upright plants to 6-10 in., with long underground stems

Pipsissewa—*C. umbellata cisatlantica*   July-Aug.   N.S. to Minn., s. to Ga., n.e. Ill. & Wis.
(King's Cure; Wintergreen; Prince's Pine)

Striped Pipsissewa—*C. maculata*   June-Aug.   s. N.H. to s. Ont., s. Mich. & n.e. Ill., s. to Ga., Ala. & Tenn.
(Spotted Wintergreen; Spotted Ratbane)

flowers typical of both species; white, sometimes pink

fruit typical of both species

Striped Pipsissewa

Pipsissewa

MP 170

GC

PACHYSANDRA—*Pachysandra terminalis*
(Japanese Spurge)

Monoecious   April-May

Japan; escaped elsewhere   Mat-forming; 10-12 in.

Note: Allegheny Spurge—*P. procumbens*—is a southern species, semi-evergreen, with less glossy leaves, producing flowers at the base of the stem instead of at the top, above the leaves, as seen here for Japanese Spurge.

Note: The top flowers of the cluster are male (staminate). The bottom ones (two visible here) are female (pistillate). The male flowers soon disintegrate, leaving only female ones which develop into the white berry-like fruit shown below.

fruit shown not fully developed (rounder, and soft, when mature)

PERIWINKLE—*Vinca minor*
(Myrtle)

Perfect   April-Sept.

Europe & w. Asia; escaped elsewhere
Long creeping stems; mat-forming

flowers blue (white or purple in variety)

fruit

Note: There is also a form with variegated leaves.

MP 171

GC

337

VINES

The plants in this section are large woody plants which, under normal circumstances, never form upright shrubs. They may climb by means of tendrils or aerial rootlets attached to some type of support, or by twining around the support. If no support is available, the climbing plants can only trail along the ground, sometimes forming dense mats. Several plants that are not climbers never take an upright form, but trail along the ground or scramble over low objects. They are included here as they have long woody stems unlike the slender growth of somewhat similar plants in the Ground Cover section. A perusal of both sections will make the distinction clear.

Some vines belong to genera that also include upright shrubs. In such cases, a cross reference is made to the Master Page that gives both types together. In all other cases, the vines are shown here only.

Rambler Roses and the so-called Matrimony-vine are not included here, as they are not true vines. They sometimes scramble over other objects, but do not really climb. If no support is available, they become upright, arching shrubs and are never found prostrate or trailing along the ground. They will be found in the Keys and Master Pages for Broad-leaved Upright Shrubs.

ARRANGEMENT OF VINES

I Alternate Leaves and Buds

    (1) Vines climbing by tendrils   MP 173-181   Ampelopsis
        Tendrils grow only at the nodes   Grape
        (joints)   Woodbine
                                                         Boston Ivy
                                                         Virginia Creeper
                                       Green-Brier or Cat-Brier

    (2) Vines climbing by aerial rootlets   MP 182-183   English Ivy
        Rootlets are not confined to nodes,   Poison Ivy
        but appear along whole sections
        of stem

    (3) Vines climbing by twining   MP 184-191   Moonseed
        No tendrils or rootlets   Dutchman's Pipe
                                               Bittersweet
                                               Wisteria
                                              Nightshade

    (4) Vines trailing or scrambling   MP 192-193   Currant
                                                Dewberry

II Opposite Leaves and Buds   MP 194-200   Clematis
    As there are only four genera, no sub-   Eunonymus
    divisions are given   Honeysuckle
                                               Trumpet-vine

HEARTLEAF AMPELOPSIS—*Ampelopsis cordata*   Perfect   May-June   High-climbing
Va., Ohio, Ill. & s.e. Neb., s. to Fla., Tex. & Mexico; naturalized n. to Mass.

Note: Pepper-vine or Thunderberry—*A. arborea*—is a southern vine with compound leaves.

Porcelainberry—*A. brevipedunculata*—from Asia, but occasionally escaped, has leaves with three to five lobes.

flowers open briefly, and petals fall almost at once

berries pink or lilac, changing to blue

Stems have white pith (Grapes have brown pith; see MP 174).

Note: These tendrils do not have adhesive disks (see Virginia Creeper, MP 179).

Note: The leaves (and buds) are alternate, but tendrils, when present, are opposite the leaves (or buds).

MP 173

V

GRAPE—*Vitis*   Dioecious or Polygamous   Mostly high-climbing

Fox Grape	*V. labrusca*	May-June	N.E. to Mich., s. to Ga., Tenn. & Ill.

(This is the parent of the Concord, Catawba, Chautauqua and many other cultivated grapes.)

Muscadine Grape (Scuppernong)	*V. rotundifolia*	June	Del. to Fla., w. to Mo., Okla. & Tex.
Riverbank Grape	*V. riparia*	May-June	N.S., N.B. & Man., s. to Va., Mo., Tex. & N. Mex.
Summer Grape	*V. aestivalis*	May-June	Vt. to Mich. & Wis., s. to Ga. & Tex.

(The Silverleaf Grape—*V. a. argentifolia*—a variety of the Summer Grape, extends its range to N.H. & Minn. Rehder considered the Silverleaf Grape to be a distinct species: *V. argentifolia*; but it is now usually treated as a variety as given here.)

All Grapes have brown pith.

woody diaphragm

Most Grapes are high-climbers, have shreddy, peeling bark, stems with pith separated at the nodes by a solid woody diaphragm. They usually have tendrils forking into two or more parts toward the ends. There is one marked exception to this, the Muscadine Grape, which has relatively smooth bark, continuous pith and tendrils that do not fork. (The Bush Grape—*V. rupestris*—has few, if any, tendrils and is not a high-climber, but in other respects it is similar to other Grapes.)

There are a good number of Grape species, which tend to be confusing because of the small differences between them; in addition, considerable hybridization takes place and intermediate forms are common. For this reason, only a few main types are given here to represent the genus.

typical
Grape flowers

Fox Grape
dark blue, black, red, brown
or green, usually with bloom

(Muscadine G. similar;
dull purple or bronze;
no bloom; very tough skin)

Summer Grape
dark blue to black,
usually with bloom

(Riverbank G. similar,
usually with heavier bloom)

MP
174

V

340

tendrils
forked

small stems
without dots

brown pith,
separated at
nodes (joints)

Summer Grape
(Fox G. and Riverbank G. similar,
but see distinctions below)

tendrils
unforked

small stems
covered with dots

brown pith,
continuous

Muscadine Grape

(1) **Fox Grape**
tendrils (or flowers) opposite
at least three consecutive
leaves (or buds)

(2) **Riverbank G. and Summer G.**
(also most other Grapes)
tendrils (or flowers) usually
missing opposite at least every
third leaf (or bud)

Note: The leaves and buds of
all Grapes are alternate, but
the tendrils are opposite the
leaves or buds.

MP
175

V

The leaves on this page are rusty or pale beneath.

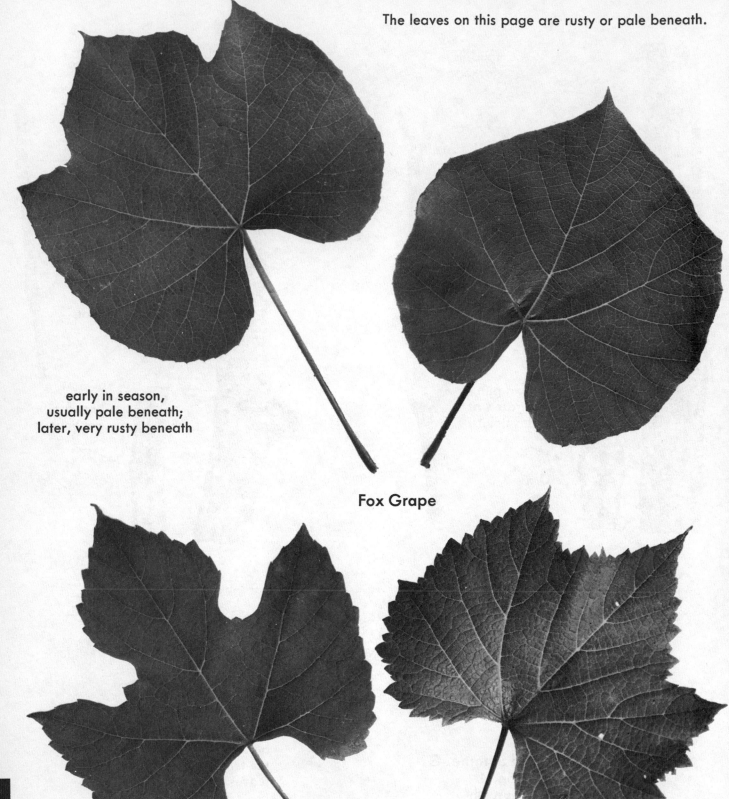

early in season,
usually pale beneath;
later, very rusty beneath

Fox Grape

pale at first;
later in season,
often somewhat
rusty beneath

MP
176

V

Summer Grape
(Silverleaf G. similar, but very white and smooth beneath)

342

The leaves on this page are green beneath.

Riverbank Grape

MP
177

V

Muscadine Grape

WOODBINE—*Parthenocissus*

	Perfect (rarely Polygamous)	High-climbing	
Boston Ivy	*P. tricuspidata*	June-Aug.	Japan, China; locally escaped
Virginia Creeper (Woodbine; Five-leaved Ivy)	*P. quinquefolia*	June-Aug.	Me. & Que. to Minn., s. to Fla., Tex. & Mexico

berries blue to black with slight bloom, borne on red stalks

Virginia Creeper

berries blue to black with bloom

Boston Ivy

MP 178

V

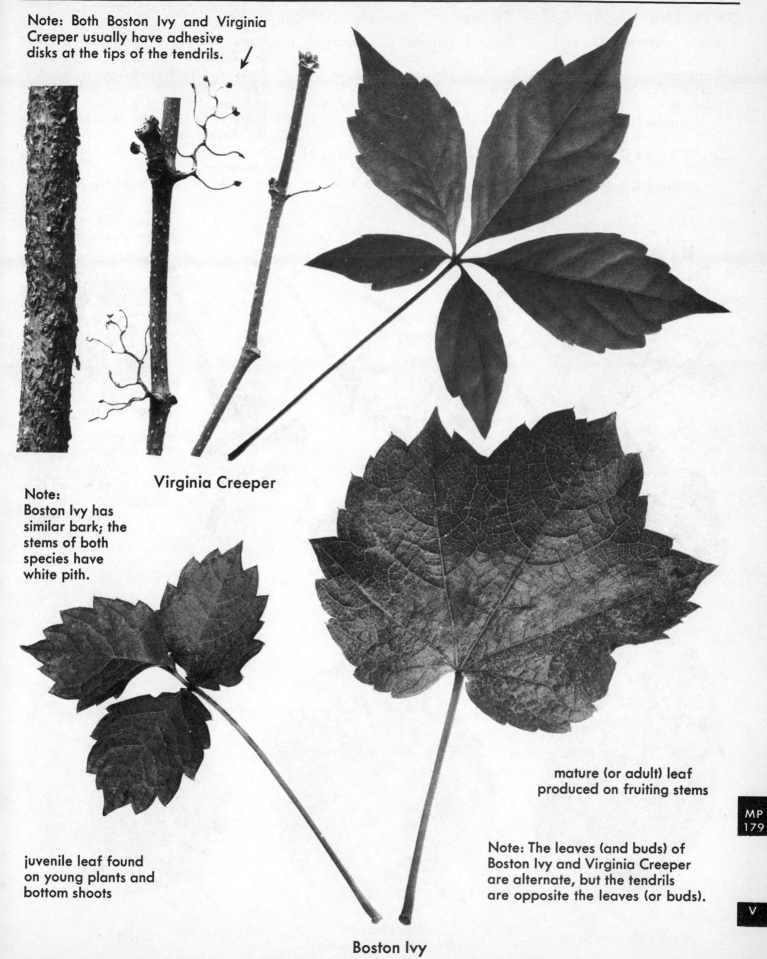

Note: Both Boston Ivy and Virginia Creeper usually have adhesive disks at the tips of the tendrils.

Virginia Creeper

Note: Boston Ivy has similar bark; the stems of both species have white pith.

mature (or adult) leaf produced on fruiting stems

juvenile leaf found on young plants and bottom shoots

Note: The leaves (and buds) of Boston Ivy and Virginia Creeper are alternate, but the tendrils are opposite the leaves (or buds).

MP 179

V

Boston Ivy

GREEN-BRIER or CAT-BRIER — *Smilax*   Dioecious   High-climbing

Bristly (or Hispid) Green-Brier	*S. hispida*	June	Conn. to Ont., Minn. & S. Dak., s. to Ga. & Tex.
Cat-Brier (Saw-Brier; Glaucous Green-Brier)	*S. glauca*	June	Mass. & s.e. N.Y. to Mo., s. to Fla. & Tex.
Horse-Brier (Common G.; Bull-Brier; Round-leaved Brier)	*S. rotundifolia*	May-June	N.S. near coast to s.e. N.H., w. from Mass. to N.Y., s. Ont. & s. Mich., s. to Fla. & Tex.
Laurel-leaved Green-Brier	*S. laurifolia*	July-Sept.	N.J., Va. & Tenn., s. to Fla. & Tex.; evergreen
Red-berried Green-Brier	*S. walteri*	June	N.J., Va. & Tenn., s. to Fla. & La.

typical Green-Brier flowers

female (pistillate)

male (staminate)

berries blue to black, sometimes with bloom

Horse-Brier

similar:
Bristly G.
Cat-Brier
Laurel-leaved G.
similar, but with red berries: Red-berried G.

evergreen

Laurel-leaved Green-Brier

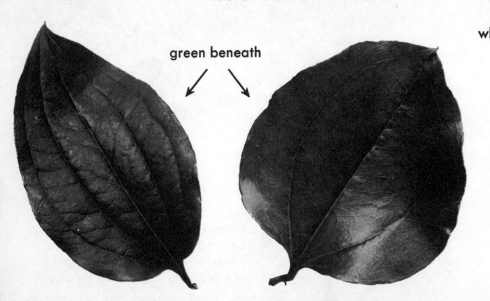

green beneath

whitish beneath when fully developed

Bristly Green-Brier   **Horse-Brier** (Red-berried G. similar)   **Cat-Brier**

MP 180

V.

thorns and bristles distinctly blackish

Bristly Green-Brier

thorns mostly green

thorns
at nodes
(joints)
as well as
in between

stems green,
usually with
some bloom

Cat-Brier

no thorns
at nodes
(joints)

stems green,
without
bloom

Horse-Brier (thorns along entire length)
(Laurel-leaved G. and Red-berried G. similar, except that they produce thorns only on the lower
parts of the stems)

MP
181

V

ENGLISH IVY — *Hedera helix*   Perfect   Aug.-Oct.   Evergreen
Europe & e. Asia; escaped   High-climbing

dark blue
to black

Note aerial rootlets
used in climbing.

leaf from
mature
(or adult)
form of
plant

Note: English Ivy
may not mature for
many years, and
the juvenile form is
the one commonly
seen. The flowers
and fruit are rarely
seen, as only the
adult form can pro-
duce them.

leaves from
juvenile form
of plant

POISON IVY — *Rhus radicans*   (see MP 74: Sumac)   Dioecious   June-July
N.S. to Minn., s. to Fla. & Ariz.   High-climbing

male (staminate) flowers
whitish

female (pistillate) flowers
greenish

fruit dull
white

aerial rootlets not always
evident when not climbing

climbs by means
of aerial rootlets

MP
183

V

COMMON MOONSEED — *Menispermum canadense* Dioecious May-June
w. Que. & w. N.E. to Man., s. to Ga. & Okla.   High-climbing

female (pistillate) flowers
pale green

male (staminate)
flowers pale yellow
to whitish

berries dark
blue with bloom

"moon" seed

Note two buds, one
above the other.

Note: The leaf-stalk is attached
inside the edge of the leaf on
the underside.

MP
184

V

DUTCHMAN'S PIPE — *Aristolochia durior*

Perfect   May   w. Pa., s. to Ga. & Ala.;
widely planted & naturalized outside this area   High-climbing

cross
section of
stem

small stems green

clusters
of side twigs
often found

MP
185

V

BITTERSWEET — *Celastrus*   Dioecious (rarely Polygamous)   High-climbing

American Bittersweet (Waxwork)	C. *scandens*	May-June	Me. & Que. to Man., s. to Ga., La. & Okla.
Oriental Bittersweet	C. *orbiculata*	May-June	e. Asia; much planted and commonly escaped

outside case orange;
berry within, red

very subject to scale

American Bittersweet

Note:
Flowers and fruit of
American Bittersweet
are terminal
(at ends of stems only);
those of Oriental B.
are axillary
(along the stems).

outside case yellow;
berry within, red

MP
186

V

Oriental Bittersweet

American Bittersweet

typical Bittersweet bark   **Oriental Bittersweet**

MP
187

V

353

WISTERIA — *Wisteria*   usually Perfect   High-climbing

American Wisteria	*W. frutescens*	May-July	Va. to Fla. & Ala.

(*W. macrostachya*, sometimes considered a variety of American Wisteria, has larger flower clusters and is found in a more westerly range: Ky. to Mo., s. to La. & Tex.)

Chinese Wisteria	*W. sinensis*	May-June	China; much planted and escaped
Japanese Wisteria	*W. floribunda*	May-June	Japan; much planted and escaped

lavender flowers, but varieties and hybrids range from white to rose and purple

MP 188

V

smooth pods →

velvety pods →

Japanese Wisteria

Note: Flowers of Japanese Wisteria open gradually, those at the tip last. Flowers of Chinese W. all open at approximately the same time; otherwise they are similar.

pods typical of both Chinese and Japanese Wisteria

American Wisteria

spurs typical
of all
Wisterias

bark grayish brown throughout

barks typical of Chinese and Japanese Wisteria

bark dark, somewhat glossy; dots (lenticels) evident; this size nearly orange

Note:
The stems of
all Wisterias
can become
very large,
but the sizes
shown here
are the most
distinctive.

MP
189

V

American Wisteria

American Wisteria
leaflets:
nine to fifteen

Chinese Wisteria
leaflets: seven to thirteen
(Japanese W.
leaflets: thirteen to nineteen)

MP
190

V

NIGHTSHADE — *Solanum dulcamara*   Perfect June-Aug.   Europe, n. Africa, e. Asia;
(Bitter-sweet)   naturalized in much of N. Amer. Trailing or occ. twining to 8-10 ft.

flowers purple
with yellow centers

It is well worth while
using a magnifying
glass to see the details
of this flower.

berries red
when ripe

The stems and leaves have
a distinctly unpleasant
odor when crushed.

MP
191

V

357

CURRANT — *Ribes*   (see MP 30)   Perfect

Skunk Currant	*R. glandulosum*	May-Aug.	Lab. to B.C., s. to s. N.E., Pa., W. Va., mts. of N.C., n. Ohio, Mich., Wis. & Minn. Most parts of this plant have a strong skunk odor when crushed. Stems trailing, with ascending branches to 2 ft.
Swamp Red Currant	*R. triste*	April-June	Lab. to Alaska, s. to s. N.E., N.Y., Pa., W. Va., Ill., Ia. & west. Low arching or creeping stems

flowers erect; white, with pinkish tinge

flowers drooping; salmon or dull pink

bristly red berries

smooth red berries

red buds

reddish leaf-stalk

Skunk Currant

Swamp Red Currant

MP
192

V

DEWBERRY — *Rubus*   (see MP 46)   Perfect   Prostrate trailing, or scrambling over low objects

| Common Dewberry | *R. flagellaris* | May-June | Me. to Man., s. to Va. & Mo.; deciduous |
| Swamp Dewberry | *R. hispidus* | June-July | N.S. to Mich., s. to Ga. & Ill.; semi-evergreen |

typical Dewberry flowers

fruit sweet;
black when ripe

stems with stoutish
prickles; not bristly

fruit sour;
black when
ripe

bristly
stems

MP
193

Swamp Dewberry   Common Dewberry

CLEMATIS — *Clematis*   Perfect (rarely Dioecious)

Purple Clematis	*C. verticillaris*	May-June	Me. & Que. to Man., s. to Md., W. Va., Ohio, Mich. & n.e. Ia. Climbs to 10 ft.
Sweet Autumn Clematis	*C. paniculata*	Aug.-Oct.	Japan; much planted & escaped here Climbs to 30 ft.
Virgin's Bower *C. virginiana* (Traveler's Joy; Love Vine)		July-Sept.	N.S. to Man., s. to Ga., La. & Kans. Climbs to 20 ft.

pale lavender
to purple
flowers

Purple Clematis

white
flowers

Virgin's Bower

white
flowers

Sweet Autumn Clematis

Typical Clematis bark and twigs, showing remnants of leaf-stalks used as tendrils for climbing.

MP
195

V

CLEMATIS (con't.)

Note: Clematis climbs by using its leaf-stalks to grip available supports, instead of having separate tendrils.

Note: Purple Clematis sometimes has no teeth along edges of leaflets.

The leaves of Sweet Autumn Clematis almost never have teeth.

The leaves of Virgin's Bower almost always have teeth.

Purple Clematis

Sweet Autumn Clematis

MP
196

V

Virgin's Bower

EUONYMUS — *Euonymus* (see MP 80)   usually Perfect

Running Strawberry-bush   *E. obovata*   May   w. N.Y., s. Ont. & Mich., s. to Md., Tenn. & Mo.
Trailing stems; branches ascending to 12 in.

Wintercreeper   *E. fortunei*   June-July   China; evergreen
   More commonly found here in the following varieties:
   Common (or Small-leaved) W.   *E. f. radicans*   Climbing, trailing and mat-forming
   Bigleaf W.   *E. f. vegeta*   Climbing, trailing or somewhat bushy; heavier
   than Common W.

typical Wintercreeper flowers; greenish white

outside case of fruit white; seed covering orange

Bigleaf Wintercreeper

Note rootlets used by Wintercreepers for climbing.

Common Wintercreeper

outside case of fruit pinkish red; seed covering orange

flowers greenish purple

MP 197

V

Running Strawberry-bush

HONEYSUCKLE — *Lonicera*   (see MP 146)   Perfect   Climbs by twining

Hairy Honeysuckle   *L. hirsuta*   May-July   Que. to Sask., s. to w. Vt., N.Y., Pa., Ohio, Mich., Minn. & Neb.   High-climbing

Japanese Honeysuckle   *L. japonica*   June-Sept.   Asia; much planted & commonly escaped here; semi-evergreen   High-climbing, trailing and mat-forming

Limber Honeysuckle   *L. dioica*   May-June   Me. & Que. to Sask., s. to Ga., e. Tenn., n.e.
(Glaucous H., Mountain H., Smoothed-leaved H.)   Ill. & Ia., in var. to Kans.   Slightly climbing or bushy

Trumpet Honeysuckle   *L. sempervirens*   May-Sept.   Me. to Mich. & Ia., s. to Fla. & Tex.; semi-evergreen   High-climbing

flowers coral red

berries red

black berries

flowers white, turning yellow

Trumpet Honeysuckle

Japanese Honeysuckle

red berries

flowers pale yellow (sometimes turning orange)

Note: All Honeysuckle vines have hollow stems.

Limber Honeysuckle
(Hairy H. similar)

Limber Honeysuckle
(Hairy H. similar, but slightly larger and very deep yellow)

barks typical of vine Honeysuckles

uppermost leaves connate
(two leaves joined together)

uppermost leaves
not connate

leaves
whitish
beneath

Note:
There is a
variety with
yellow-veined leaves.

Japanese Honeysuckle

cut-leaf variety
of Japanese H.
occasionally
encountered

Trumpet Honeysuckle
(high-climbing)
(Limber H. similar;
slightly climbing or bushy)

upper leaves connate;
leaves green and hairy on both sides;
hairs especially evident on leaf edges

MP
199

V

Hairy Honeysuckle

**Hairy
Honeysuckle**

**Limber
Honeysuckle**

TRUMPET-VINE — *Campsis radicans*   Perfect   July-Sept.   Pa. to la., s. to Fla. & Tex.;
(Trumpet-creeper; Trumpet Flower; Cow-itch)   naturalized n. to Conn. & Mich.
                                                                            High-climbing

flowers orange red

Note rootlets used
in climbing.

MP
200

V

BIBLIOGRAPHY

AMMONS, NELLE. *Shrubs of West Virginia.* (Herbarium, Contribution No. 55.) Morgantown: West Virginia University, 1950.

BILLINGTON, CECIL. *Shrubs of Michigan.* 2nd ed. Bloomfield Hills, Mich.: Cranbrook Institute of Science, 1949.

BLAKE, S. F. *Guide to Popular Floras of the United States and Alaska.* Bibliographical Bulletin No. 23, U.S. Department of Agriculture, 1954.

CORE, EARL L. and AMMONS, NELLE. *Woody Plants in Winter.* Pittsburgh: Boxwood Press, 1958.

DEAM, CHARLES C. *Shrubs of Indiana.* (Indiana Conservation Department Publication No. 44.) Indianapolis: W. B. Burford, 1924.

DOLE, E. J. *The Flora of Vermont.* 3rd rev. ed. Burlington, Vt.: Free Printing Press, 1937.

FASSETT, NORMAN C. *Spring Flora of Wisconsin.* 3rd ed. Madison: University of Wisconsin Press, 1957.

FERNALD, M. L. *Gray's Manual of Botany.* 8th ed. New York: American Book Co., 1950.

FREEMAN, OLIVER M. *Annotated List of the Plants Growing Naturally at the National Arboretum.* (National Arboretum Contribution No. 1.) Washington, D.C.: Government Printing Office, 1953.

GRAVES, ARTHUR HARMOUNT. *Illustrated Guide to Trees and Shrubs.* Wallingford, Conn.: published by the author, 1952.

GRIMM, WILLIAM CAREY. *The Shrubs of Pennsylvania.* Harrisburg, Pa.: Stackpole Co., 1952.

HARRINGTON, HAROLD D. *Keys to the Woody Plants of Iowa.* University of Iowa Studies in Natural History, Vol. 17, No. 9. Iowa City: The University, 1940.

HERMANN, FREDERICK J. *A Checklist of Plants in the Washington-Baltimore Area.* 2nd ed. Washington, D.C.: Smithsonian Institution, 1946.

HODGDON, ALBION R. and STEELE, FREDERIC L. *The Woody Plants of New Hampshire.* (Bulletin 447, Agricultural Experiment Station.) Durham, N.H.: University of New Hampshire, 1958.

HOUSE, HOMER D. *Annotated List of the Ferns and Flowering Plants of New York State.* Albany: University of the State of New York, 1924.

HYLAND, FAY and STEINMETZ, FERDINAND H. *The Woody Plants of Maine.* Orono: University of Maine Press, 1944.

MASSEY, ARTHUR B. *Evergreen Shrubs, Vines and Ground Cover Native in Virginia.* Blacksburg: Virginia Polytechnic Institute, 1952.

REHDER, ALFRED. *Manual of Cultivated Trees and Shrubs*. 2nd ed. New York: Macmillan Co., 1940.

ROSENDAHL, CARL OTTO. *Trees and Shrubs of the Upper Midwest*. Minneapolis: University of Minnesota Press, 1955.

SHANKS, ROYAL E. *Checklist of Woody Plants of Tennessee*. Knoxville: University of Tennessee, 1952.

STONE, WITMER. *The Plants of Southern New Jersey*. Trenton: 1911.

TEHON, LEO R. *Fieldbook of Native Illinois Shrubs*. Urbana: Natural History Survey Division, State of Illinois, 1942.

GLOSSARY

Appressed—close to (not sticking out); opposite of divergent.

Axillary—(as buds) borne in angle or axil formed by two diverging parts.

Bloom—whitish, usually powdery, covering that can be rubbed off.

Bract—a modified leaf, usually at the base of flowers or leaves.

Calyx—outer part of a flower (at the bottom when the flower is open). It is usually green (but occasionally colored, especially when a flower produces no colored petals).

Compound—having two or more similar parts; a leaf made up of several leaflets on the same leaf-stalk is a compound leaf.

Deciduous—not permanent, falling off, as with leaves in autumn.

Dioecious—male flower on one plant, female on another.

Divergent—sticking out; opposite of appressed.

Genus—main name of a plant (Dogwood, Honeysuckle, etc.).

Hybrid—a cross between closely related plants.

Leaf-scar—a scar marking the point of separation from the twig when a leaf drops off in the fall.

Leaflets—the single blades of a compound leaf.

Lenticels—openings on the surface of leaves or bark allowing the passage of air or gases; these appear as dots on the surface.

Lobe—the part of a leaf sticking out when the margins are not uniform. This is the opposite of the sinus, which is the cut-out part of an uneven leaf.

Monoecious—male and female flowers distinct from one another, but on the same plant.

Mucro—short, abrupt tip.

Node—the point on the twig that bears a leaf or leaves.

Palmate—spreading from one point; a palmately compound leaf is one where all the leaflets branch from one point.

Perfect—both male and female parts in the same flower.

Petiole—leaf-stalk.

Pinnate—having parts arranged along two sides; a pinnately compound leaf has leaflets along both sides of the leaf-stalk.

Pistil—the seed-bearing (female) part of a flower.

Pistillate flower—female flower.

Pith—inner part of a twig or stem that is not woody but pulpy.

Polygamous—bearing both perfect and unisexual flowers on the same plant.

Raceme—elongated flower (or fruit) cluster, the individual flowers (or fruit) arranged along a single axis or stem.

Rachis—the continuation of the leaf-stalk of a compound leaf, beginning at the lowest leaflet(s) and continuing to the end leaflet.

Revolute—rolled backward from the margins or the apex.

Sepals—divisions of the calyx (sometimes petal-like).

Sessile—without a stalk.

Simple—applied to a leaf with only one blade.

Sinus—the cut-out part of a leaf with uneven margins; the opposite of lobe.

Species—a particular kind of plant within a given genus, such as Flowering Dogwood (Dogwood being the genus).

Stamen—the pollen-bearing (male) part of a flower.

Staminate flower—male flower.

Sterile—incapable of bearing fruit.

Stipules—leaf-like objects at the base of leaf-stalks.

Sub—prefix indicating "somewhat," "slightly" or "partly."

Subgenus—a subdivision of a genus.

Subshrub—a term used to indicate a plant intermediate between a perennial and a true woody shrub. The base is always woody.

Whorl—a circle of three or more similar parts around a central point, as three or more leaves growing around a twig from one joint, or node.

INDEX
TO
MASTER PAGES

INDEX TO MASTER PAGES

Alder, 6
  American Green
  Black (Holly), 78
  Mountain
  Sea-side
  Smooth
  Speckled
  White (Clethra, 102)

Allspice, Carolina
  (Sweet Shrub), 20

Alnus, 6
  A. crispa
  A. incana
  A. maritima
  A. rugosa

Althea, Shrub (Rose of Sharon),
90

Amelanchier, 44

Amorpha, 66
  A. canescens
  A. fruticosa

Ampelopsis, Heartleaf, 173

Ampelopsis, 173
  A. arborea
  A. brevipedunculata
  A. cordata

Andromeda, 114
  Japanese
  Mountain

Andromeda glaucophylla, 113A

Angelica-tree (Sarsaparilla), 92

Aralia, 92
  A. hispida
  A. spinosa

Arbutus, Trailing, 169B

Arctostaphylos uva-ursi, 164A

Arctous alpinus, 166B

Aristolochia durior, 185

Aronia, 39
  A. arbutifolia
  A. melanocarpa
  A. prunifolia

Arrow-wood (Viburnum), 136

Ascyrum, 89A
  A. hypericoides
  A. stans

Ash, Prickly, 70

Ash, Wafer-
  (Hop-tree), 71

Azalea, 104
  Alpine (Alpine Azalea, 167C)
  Clammy
  Flame
  Honeysuckle
  Mountain
  Pink
  Rose
  Swamp

Baccharis halimifolia, 155

Barberry, 18
  American
  Common
  Japanese

Bayberry, 4
  Evergreen

Beach Heather, 161C
  Golden
  Woolly

Bearberry, Alpine, 166B

Bearberry, Common, 164A

Beech, Blue (Hornbeam), 12

Benjamin Bush (Spice-bush), 21

Berberis, 18
  B. canadensis
  B. thunbergii
  B. vulgaris

Betula, 9
  B. glandulosa
  B. pumila

Bilberry (Blueberry), 119
  Bog
  Dwarf

Birch, 9
  Dwarf
  Low

Bittersweet, 186
  American
  Oriental

Bitter-sweet (Nightshade), 191

Black Alder (Holly), 78

Blackberry, 46
  Allegheny
  Canada
  Highbush
  Smooth
  Thornless

Black-cap (Raspberry), 46

Black-haw (Viburnum), 136

Bladdernut, American, 87

Blueberry, 119
  Black Highbush
  (Blue) Highbush
  Dwarf Sweet
  Early Lowbush
  Late Lowbush
  Low Sweet

Bramble, 46

Broom, Scotch, 65B

Broom Crowberry, 160A

Buckbush (Coralberry), 145

Buckhorn (Clubmoss), 162

Buckthorn, 84
  Alder (European)
  Alder-leaved
  Common
  European
  Lance-leaved

Buffalo-berry, 95
  Canada

Buffalo-nut, 16

Bull-Brier (Green-Brier), 180

Bunchberry (Dogwood), 98

Burning-bush (Euonymus), 80

Bursting Heart (Euonymus), 80

Bush-honeysuckle, 154

Buttonbush, 144

Calluna vulgaris, 160D

Calycanthus, 20
  C. fertilis
  C. floridus

Campsis radicans, 200

Carolina Allspice (Sweet Shrub), 20

Carpinus caroliniana, 12

Cassandra, Dwarf (Leatherleaf), 116B

Cassiope, 161B

Cassiope hypnoides, 161B

Castanea pumila, 13

Cat-Brier, 180

Ceanothus americanus, 45A

Celastrus, 186
  C. orbiculata
  C. scandens

Cephalanthus occidentalis, 144

Cercis canadensis, 68

Chamaedaphne calyculata, 116B

Checkerberry (Wintergreen), 170A

Cherry, 54
  Choke
  Sand

Chestnut, 13

Chimaphila, 170B
  C. maculata
  C. umbellata cisatlantica

Chinquapin (Chestnut), 13

Chinquapin Oak, 14

Chiogenes hispidula, 168C

Chionanthus virginicus, 129

Chittam-wood (Smoke-tree) 72

Chokeberry, 39
  Black
  Purple
  Red

Cinquefoil, 64
  Shrubby
  Three-toothed

Clematis, 194
  Purple
  Sweet Autumn

Clematis, 194
  C. paniculata
  C. verticillaris
  C. virginiana

Clethra, 102

Clethra alnifolia, 102

Cliff Green (Pachistima), 164B

Clubmoss, 162
  Running
  Shining
  Tree

Comptonia peregrina, 11B

Consumption-weed (Groundsel-bush), 155

Coralberry, 145

Corema conradii, 160A

Cornel (Dogwood), 98
  Dwarf (Bunchberry)
  Swamp

Cornus, 98
  C. alternifolia
  C. amomum
  C. canadensis
  C. florida
  C. obliqua
  C. racemosa
  C. rugosa
  C. stolonifera

Corylus, 10
  C. americana
  C. cornuta

Cotinus, 72
  C. americanus
  C. coggygria

Cowberry (Blueberry), 119
  (Mount Cranberry)

Cow-itch (Trumpet-vine), 200

Cranberry (Blueberry), 119
  Large
  Mountain
  Rock
  Small

Cranberry, Highbush (Viburnum), 136

Crataegus, 43

Creeper, Trumpet- (Trumpet-vine), 200

Creeper, Virginia (Woodbine), 178

Creeping Jenny (Clubmoss), 162

Creeping Savin (Juniper), 159B

Creeping Snowberry, 168C

Crowberry, 160B

Crowberry, Broom, 160A

Curleberry (Crowberry), 160B

Currant, 30
  American Black
  Bristly Black
  European Black
  Garden Red
  Indian (Coralberry, 145)
  Skunk
  Swamp Red
  Wild Black

Cytisus scoparius, 65B

Dangleberry (Huckleberry), 126

Daphne, 91B

Daphne mezereum, 91B

Decodon verticillatus, 97B

Deerberry (Blueberry), 119

Deutzia, 26

Deutzia scabra, 26

Devil's Walking-stick (Sarsaparilla), 92

Dewberry, 46
  Common
  Swamp

Diapensia, 167D

Diapensia lapponica, 167D

Diervilla lonicera, 154

Dirca palustris, 91A

Dockmackie (Viburnum), 136
    (Maple-leaved Viburnum)

Dog Hobble (Leucothoë), 118

Dogwood, 98
    Alternate-leaved
    Dwarf
    Flowering
    Gray
    Gray-stemmed
    Panicled (Gray-stemmed)
    Red-Osier
    Red-stemmed (Red-Osier)
    Round-leaved
    Silky (Swamp)
    Swamp

Dutchman's Pipe, 185

Dyer's Greenweed, 65A

Eglantine (Rose), 57

Elaeagnus, 96
    Autumn

Elaeagnus, 96
    E. angustifolia
    E. commutata
    E. umbellata

Elder, 134
    American
    Common
    Dwarf (Sarsaparilla, 92)
    Marsh- (Marsh-elder, 156)
    Red-berried
    Scarlet

Elderberry, 134

Election Pink (Azalea), 104

Empetrum nigrum, 160B

Epigaea repens, 169B

Erica, 160C
    E. carnea
    E. vagans

Euonymus, 80
    Brook
    Running
    Winged

Euonymus, 80
    E. alata
    E. americana
    E. atropurpurea
    E. europaea
    E. fortunei
        E. f. radicans
        E. f. vegeta
    E. obovata

Evergreen (Clubmoss), 162
    Coral
    Running
    Staghorn
    Trailing

False Indigo, 66

False Spirea, 24

Fetter-bush (Andromeda), 114

Fetter-bush (Leucothoë), 118

Firethorn, 42

Flowering Moss, 165C

Forsythia, 131
    Border
    Greenstem
    Weeping

Forsythia, 131
    F. intermedia
    F. suspensa
    F. viridissima

Fringe-tree, 129

Gale, Sweet, 4

Gaultheria procumbens, 170A

Gaylussacia, 126
    G. baccata
    G. brachycera
    G. dumosa
    G. frondosa

Genista tinctoria, 65A

Globeflower (Buttonbush), 144

Golden-bells (Forsythia), 131

Gooseberry, 30
    European
    Low Wild
    Prickly
    Round-leaved (see note under
    Smooth Gooseberry in text)
    Smooth

Gorse, 67B

Grape, 174
    Bush
    Fox
    Muscadine
    Riverbank
    Silverleaf
    Summer

Green-Brier, 180
    Bristly
    Common
    Glaucous
    Hispid
    Laurel-leaved
    Red-berried

Ground Cedar (Clubmoss), 162

Ground Hemlock (Yew), 159A

Ground Pine (Clubmoss), 162

Groundsel-bush, 155

Halesia carolina, 128

Hamamelis virginiana, 38

Hardhack (Spirea), 40

Hawthorn, 43

Hazel (Hazelnut), 10

Hazel, Witch-, 38

Hearts-a-burstin (Euonymus),
    80

Heath, 160C
    Cornish
    Spring

Heath, Mountain, 161A

Heather, 160D

Heather, Beach,
    see Beach Heather

Hedera helix, 182

Hercules' Club (Sarsaparilla), 92

Hibiscus syriacus, 90

Highbush Cranberry
    (Viburnum), 136

Hills of Snow (Hydrangea), 28
    (var. of Wild Hydrangea)

Hobble-bush (Viburnum), 136

Holly, 78
    Large-leaved
    Mountain (Large-leaved)

Holly, Mountain, 83

Honeyballs (Buttonbush), 144

Honeysuckle, 146
American Fly-
Amur
Blue (Mountain Fly-)
Bush- (Bush-honeysuckle,
154)
European Fly-
Glaucous (Limber)
Hairy
Japanese
Limber
Morrow
Mountain (Limber)
Mountain Fly-
Smooth-leaved (Limber)
Swamp (Azalea, 104)
Swamp Fly-
Ta(r)tarian
Trumpet
Wild (Azalea, 104)

Hop-tree, 71

Hornbeam, American, 12

Horse-Brier (Green-Brier), 180

Huckleberry, 126
Black
Blue
Box-
Dwarf
Evergreen
Highbush
Squaw (Blueberry, 119)
Swamp

Hudsonia, 161C
H. ericoides
H. tomentosa

Hydrangea, 28
Peegee
Wild

Hydrangea, 28
H. arborescens
H. a. grandiflora
H. paniculata grandiflora

Hyeble (Sarsaparilla), 92

Hypericum, 88
H. densiflorum
H. prolificum

Ilex, 78
I. glabra
I. montana
I. verticillata

Indian Currant (Coralberry),
145·

Indigo Bush (False Indigo), 66

Inkberry (Holly), 78

Ironwood (Hornbeam), 12

Itea virginica, 23B

Iva frutescens oraria, 156

Ivy (Laurel), 110

Ivy (Woodbine), 178
Boston
Five-leaved

Ivy, English, 182

Ivy, Poison (Sumac), 74
(*see also* Vines, 183)

Jetbead, 45B

Juneberry (Shadbush), 44

Juniper, 159B
Common
Creeping
Ground

Juniperus, 159B
J. communis
J. c. depressa
J. horizontalis

Kalmia, 110
K. angustifolia
K. latifolia
K. polifolia

King's Cure (Pipsissewa), 170B

Kinnikinic (Common
Bearberry), 164A

Knotweed, Japanese, 17B

Labrador Tea, 116A

Lamb-kill (Laurel), 110

Laurel, 110
Great (Rhododendron, 108)
Mountain-
Pale
Purple (Rhododendron, 108)
Sheep-
Swamp-

Lead-plant (False Indigo), 66

Leatherbush (Leatherwood),
91A

Leatherleaf, 116B

Leatherwood, 91A

Ledum groenlandicum, 116A

Leiophyllum buxifolium, 113B

Leucothoë, 118
Drooping
Swamp

Leucothoë, 118
L. catesbaei
L. racemosa

Ligustrum, 132
L. ibota
L. obtusifolium
L. o. regelianum
L. ovalifolium
L. vulgare

Lilac, 130
Common
Old-fashioned

Lindera benzoin, 21

Lingenberry (Blueberry), 119
(Mountain Cranberry)

Linnaea borealis americana,
168B

Locust, Bristly, 69

Loiseleuria procumbens, 167C

Lonicera, 146
L. canadensis
L. dioica
L. hirsuta
L. japonica
L. maackii
L. morrowii
L. oblongifolia
L. sempervirens
L. tatarica
L. villosa
L. xylosteum

Loose-strife, Swamp, 97B

Love Vine (Clematis), 194

Lycium, 125
L. chinense
L. halimifolium

Lycopodium, 162
L. clavatum
L. complanatum
L. lucidulum
L. obscurum
L. tristachyum

Lyonia, 117

Lyonia, 117
L. ligustrina
L. mariana

Male-berry (Lyonia), 117

Marsh-elder, 156

Matrimony-vine, 125
  Chinese
  Common

Mayflower (Trailing Arbutus),
  169B

Meadowsweet (Spirea), 40
  Birch-leaved
  Broad-leaved

Mealberry (Common Bearberry),
  164A

Menispermum canadense, 184

Menziesia, 103
  Allegheny

Menziesia pilosa, 103

Mezereum (Daphne), 91B

Minniebush (Menziesia), 103

Mistletoe, American, 17A

Mitchella repens, 168A

Mock-orange, 22
  Large-flowered
  Scentless

Moonseed, Common, 184

Moosewood (Leatherwood),
  91A

Moosewood (Viburnum), 136
  (Hobble-bush)

Moss, Flowering, 165C

Mountain Ash Spirea (False
  Spirea), 24

Mountain Heath, 161A

Mountain Lover (Pachistima),
  164B

Myrica, 4
  M. cerifera
  M. gale
  M. heterophylla
  M. pensylvanica

Myrtle (Periwinkle), 171B

Myrtle, Sand-, 113B

Myrtle, Sea (Groundsel-bush),
  155

Myrtle, Wax-, 4

Nanny-berry (Viburnum), 136

Nemopanthus mucronatus, 83

New Jersey Tea, 45A

Nightshade, 191

Ninebark, 27

Oak, 14
  Chinquapin
  Scrub

Oil-nut (Buffalo-nut), 16

Oleaster, 96

Olive, Russian (Oleaster), 96

Opossum-wood (Silverbell), 128

Pachistima, 164B

Pachistima canbyi, 164B

Pachysandra, 171A

Pachysandra, 171A
  P. procumbens
  P. terminalis

Parthenocissus, 178
  P. quinquefolia
  P. tricuspidata

Partridge-berry, 168A

Pepperbush, Sweet (Clethra),
  102

Pepper-Vine (Ampelopsis), 173

Periwinkle, 171B

Philadelphus, 22
  P. coronarius
  P. grandiflorus
  P. inodorus

Phoradendron flavescens, 17A

Phyllodoce coerulea, 161A

Physocarpus opulifolius, 27

Pieris, 114
  P. floribunda
  P. japonica

Pigeonberry (Sarsaparilla), 92

Pimbina (Viburnum), 136
  (Highbush Cranberry)

Pinxterbloom (Azalea), 104

Pipsissewa, 170B
  Striped

Plum, 54
  Beach
  Red
  Wild
  Wild Yellow

Poison Ivy (Sumac), 74

Polygonum cuspidatum, 17B

Porcelainberry (Ampelopsis),
  173

Possum-haw (Viburnum), 136
  (Smooth Withe-rod)

Potentilla, 64
  P. fruticosa
  P. tridentata

Prickly Ash, 70

Prince's Pine (Pipsissewa), 170B

Privet, 132
  California
  Common
  Ibota
  Regal

Prunus, 54
  P. americana
  P. maritima
  P. pumila
  P. virginiana

Ptelea trifoliata, 71

Pyracantha coccinea (lalandii),
  42

Pyrularia pubera, 16

Pyxidanthera barbulata, 165C

Pyxie (Flowering Moss), 165C

Quercus, 14
  Q. ilicifolia
  Q. prinoides

Raspberry, 46
  American Red
  Black
  European Red
  Purple Flowering

Rat Stripper (Pachistima), 164B

Ratbane, Spotted (Pipsissewa),
  170B

Redbud, 68

Rhamnus, 84
  R. alnifolia
  R. cathartica
  R. frangula
  R. lanceolata

Rhododendron (104); 108
  Carolina
  Catawba
  Lapland
  Mountain
  Purple
  Rosebay

Rhododendron, 104; 108
  (Azaleas), 104
  R. calendulaceum
  R. canadense
  R. nudiflorum
  R. roseum
  R. viscosum
  (Rhododendrons), 108
  R. carolinianum
  R. catawbiense
  R. lapponicum
  R. maximum

Rhodora, 104

Rhodotypos scandens, 45B

Rhus, 74
  R. aromatica
  R. copallina
  R. glabra
  R. radicans
  R. typhina
  R. vernix

Ribes, 30
  R. americanum
  R. cynosbati
  R. glandulosum
  R. grossularia
  R. hirtellum
  R. lacustre
  R. nigrum
  R. rotundifolium
  R. sativum
  R. triste

Robinia hispida, 69

Rock Rose (Rhododendron),
  108

Rosa, 57
  R. blanda
  R. carolina
  R. eglanteria
  R. multiflora
  R. palustris
  R. rugosa
  R. setigera
  R. virginiana

Rose, 57
  Carolina
  Meadow
  Multiflora
  Pasture
  Prairie
  Rock (Rhododendron, 108)
  Rugosa
  Smooth
  Swamp
  Sweet-Brier
  Virginia

Rose of Sharon, 90

Rosebay (Rhododendron), 108

Rosemary, 113A
  Bog
  Downy Bog

Round-leaved Brier
  (Green-Brier), 180

Rubus, 46
  R. allegheniensis
  R. canadensis
  R. flagellaris
  R. hispidus
  R. idaeus strigosus
  R. occidentalis
  R. odoratus
  R. phoenicolasius

Russian Olive (Oleaster), 96

St. Andrew's Cross (St. Peter's-
  wort), 89A

St. John's-wort, 88
  Dense-flowered
  Shrubby

St. Peter's-wort, 89A

Salix, 1
  S. bebbiana
  S. discolor
  S. humilis
  S. lucida
  S. sericea
  S. tristis
  S. uva-ursi

Sambucus, 134
  S. canadensis
  S. pubens

Sand-myrtle, 113B

Sarsaparilla, 92
  Bristly

Savin, Creeping (Juniper), 159B

Saw-Brier (Green-Brier), 180

Scotch Broom, 65B

Scuppernong (Grape), 174

Sea Myrtle (Groundsel-bush),
  155

Service-berry (Shadbush), 44

Seven-bark (Hydrangea), 28

Shadbush, 44

Sheep-berry (Viburnum), 136
  (Nanny-berry)

Shepherdia, 95
  S. argentea
  S. canadensis

Shittimwood (Silverbell), 128

Shoestring (False Indigo), 66

Shrub Althea (Rose of Sharon),
  90

Shrub Yellow-root, 25

Silverbell, 128

Silver-berry (Oleaster), 96

Smilax, 180
  S. glauca
  S. hispida
  S. laurifolia
  S. rotundifolia
  S. walteri

Smoke-tree, 72

Snowberry, 145

Snowberry, Creeping, 168C

Snowdrop Tree (Silverbell), 128

Soapberry (Buffalo-berry), 95

Solanum dulcamara, 191

Sorbaria sorbifolia, 24

Spice-bush, 21

Spiceweed (Spice-bush), 21

Spikenard, 92

Spindle-tree, European
  (Euonymus), 80

Spiraea, 40
  S. corymbosa
  S. latifolia
  S. tomentosa

Spirea, 40
  Corymbed

Spoonwood (Laurel), 110

Spurge (Pachysandra), 171A
  Allegheny
  Japanese

Squaw Huckleberry (Blueberry),
  119
  (Deerberry)

Stag-bush (Viburnum), 136
  (Black-haw)

Stagger-bush (Lyonia), 117

Staphylea trifolia, 87

Steeplebush (Spirea), 40

Strawberry-bush (Euonymus),
  80
  Running

Strawberry Shrub (Sweet Shrub), 20
  Hairy
  Smooth

Sumac, 74
  Dwarf
  Fragrant
  Poison
  Shining
  Smooth
  Staghorn
  Velvet

Summer Sweet (Clethra), 102

Swamp-haw (Viburnum), 136
  (Smooth Withe-rod)

Swamp Loose-strife, 97B

Sweet Bells (Leucothoë), 118

Sweet Fern, 11B

Sweet Gale, 4

Sweet-haw (Viburnum), 136

Sweet Shrub, 20

Sweet Syringa (Mock-orange), 22

Switch-ivy (Leucothoë), 118

Symphoricarpos, 145
  S. albus
  S. orbiculatus

Syringa, Sweet (Mock-orange), 22

Syringa vulgaris, 130

Tangleberry (Huckleberry), 126

Tangle-legs (Viburnum), 136
  (Hobble-bush)

Tassle-white (Virginia-willow), 23B

Taxus canadensis, 159A

Teaberry (Wintergreen), 170A

Thimbleberry (Raspberry), 46
  (Black Raspberry)
  (Purple Flowering Raspberry)

Thunderberry (Ampelopsis), 173

Toothache-tree (Prickly Ash), 70

Toothache-tree (Sarsaparilla), 92

Trailing Arbutus, 169B

Traveler's Joy (Clematis), 194

Trumpet-creeper (Trumpet-vine), 200

Trumpet Flower (Trumpet-vine), 200

Trumpet-vine, 200

Twinberry (Honeysuckle), 146
  (American Fly-Honeysuckle)

Twinflower, 168B

Ulex europaeus, 67B

Vaccinium, 119
  V. angustifolium
  V. atrococcum
  V. cespitosum
  V. corymbosum
  V. macrocarpum
  V. oxycoccus
  V. stamineum
  V. uliginosum
  V. vacillans
  V. vitis-idaea minus

Viburnum, 136
  Downy
  Maple-leaved
  Sweet (Nanny-berry)

Viburnum, 136
  V. acerifolium
  V. alnifolium
  V. cassinoides
  V. dentatum
  V. lentago
  V. nudum
  V. prunifolium
  V. pubescens
  V. rafinesquianum
  V. trilobum

Vinca minor, 171B

Virginia Creeper (Woodbine), 178

Virginia-willow, 23B

Virgin's Bower (Clematis), 194

Vitis, 174
  V. aestivalis
    V. a. argentifolia
  V. labrusca
  V. riparia
  V. rotundifolia
  V. rupestris

Wafer-Ash (Hop-tree), 71

Wahoo (Euonymus), 80

Water Willow (Swamp Loose-strife), 97B

Wax-myrtle, 4

Waxwork (Bittersweet), 186

Whin (Dyer's Greenweed), 65A

Wicky (Laurel), 110

Wicopy (Leatherwood), 91A

Wild Raisin (Viburnum), 136
  (Nanny-berry)

Willow, 1
  Beak
  Bearberry
  Dwarf Gray
  Dwarf Pussy
  Dwarf Upland
  Glaucous (Pussy)
  Gray
  Prairie (Gray)
  Pussy
  Red (Swamp Dogwood, 98)
  Shining
  Silky
  Upland (Gray)
  Virginia- (Virginia-willow, 23B)
  Water (Swamp Loose-strife, 97B)

Wineberry (Raspberry), 46

Winterberry (Holly), 78
  Mountain

Wintercreeper (Euonymus), 80
  Bigleaf
  Common

Wintergreen, 170A

Wintergreen (Pipsissewa), 170B

Wisteria, 188
  American
  Chinese
  Japanese

Wisteria, 188
  W. floribunda
  W. frutescens
  W. macrostachya
  W. sinensis

Witch-Hazel, 38

Witch-hobble (Viburnum), 136

Withe-rod (Viburnum), 136

Woad Waxen (Dyer's Greenweed), 65A

Wolf's-claws (Clubmoss), 162

Woodbine, 178

Xanthorhiza simplicissima, 25

Yellow-root, Shrub, 25

Yew, 159A
  American
  Canada

Zanthoxylum americanum, 70